AF385815

Cognitive Technologies

Editor-in-Chief

Daniel Sonntag, German Research Center for AI
DFKI, Saarbrücken, Germany

Titles in this series now included in the Thomson Reuters Book Citation Index and Scopus!

The Cognitive Technologies (CT) series is committed to the timely publishing of high-quality manuscripts that promote the development of cognitive technologies and systems on the basis of artificial intelligence, image processing and understanding, natural language processing, machine learning and human-computer interaction.

It brings together the latest developments in all areas of this multidisciplinary topic, ranging from theories and algorithms to various important applications. The intended readership includes research students and researchers in computer science, computer engineering, cognitive science, electrical engineering, data science and related fields seeking a convenient way to track the latest findings on the foundations, methodologies and key applications of cognitive technologies.

The series provides a publishing and communication platform for all cognitive technologies topics, including but not limited to these most recent examples:

- Interactive machine learning, interactive deep learning, machine teaching
- Explainability (XAI), transparency, robustness of AI and trustworthy AI
- Knowledge representation, automated reasoning, multiagent systems
- Common sense modelling, context-based interpretation, hybrid cognitive technologies
- Human-centered design, socio-technical systems, human-robot interaction, cognitive robotics
- Learning with small datasets, never-ending learning, metacognition and introspection
- Intelligent decision support systems, prediction systems and warning systems
- Special transfer topics such as CT for computational sustainability, CT in business applications and CT in mobile robotic systems

The series includes monographs, introductory and advanced textbooks, state-of-the-art collections, and handbooks. In addition, it supports publishing in Open Access mode.

Oleksandr Kuznetsov

Intelligent Systems: From Theory to Applications

Foundations, Search Algorithms, and Machine Learning

 Springer

Oleksandr Kuznetsov
Department of Theoretical and Applied Sciences
eCampus University
Via Isimbardi 10, Novedrate (CO), 22060, Italy

ISSN 1611-2482   ISSN 2197-6635   (electronic)
Cognitive Technologies
ISBN 978-3-032-00043-9   ISBN 978-3-032-00044-6   (eBook)
https://doi.org/10.1007/978-3-032-00044-6

Foreword

It is with great pleasure that I introduce this textbook, which represents both an important educational contribution and the culmination of years of collaborative research with Dr. Oleksandr Kuznetsov.

The field of artificial intelligence has seen explosive growth in recent years, yet a persistent challenge remains: bridging the gap between theoretical concepts and practical implementation. Too often, students encounter either highly abstract mathematical treatments disconnected from real-world applications, or simplified implementations that fail to convey the underlying principles. This textbook directly addresses this challenge through its unique approach combining clear theoretical explanations with comprehensive Python implementations.

Over the past 5 years, Dr. Kuznetsov and I have collaborated on numerous research projects spanning cryptography, biometric security, image processing, and machine learning applications. These collaborations have produced over 20 joint publications in prestigious journals and conferences. What makes this textbook especially valuable is that it distills the best practices and insights from our research into an accessible educational format, allowing students to benefit from approaches proven effective in real-world applications.

The book's structure reflects a thoughtful progression through the essential components of intelligent systems. It begins with fundamental concepts that provide a solid theoretical foundation, then advances through increasingly sophisticated topics:

- The introductory chapters establish a clear framework for understanding what intelligent systems are and how they function, making these complex concepts accessible even to those new to the field.
- The sections on search algorithms—from basic uninformed search to advanced heuristic methods—demonstrate how computational problems can be efficiently solved through systematic exploration of possibility spaces.
- The chapters on optimization techniques, including genetic algorithms and simulated annealing, show how nature-inspired approaches can tackle problems traditional algorithms struggle with.

- The machine learning foundations provided in the later chapters prepare students for the rapidly evolving AI landscape, with practical implementations that demystify these powerful techniques.

What truly distinguishes this book is its commitment to practical application. Each theoretical concept is immediately reinforced with Python code that students can run, modify, and extend. This hands-on approach transforms passive learning into active skill-building, allowing students to develop confidence in their ability to implement intelligent systems.

As an educator who has witnessed the struggles students face when learning these topics, I particularly appreciate the clarity of Dr. Kuznetsov's explanations. Complex ideas are broken down into digestible components without sacrificing academic rigor. The numerous exercises throughout the text provide opportunities for students to test their understanding and build their implementation skills progressively.

The importance of this educational approach cannot be overstated in today's technological landscape. As intelligent systems become increasingly embedded in our daily lives—from recommendation engines to autonomous vehicles—we need practitioners who understand not only how to use AI frameworks but also how these systems function at a fundamental level. This deeper understanding is essential for creating responsible, effective, and innovative applications.

Dr. Kuznetsov has created a resource that will serve students at various levels, from undergraduates taking their first steps into the field to graduate students seeking to deepen their practical implementation skills. I am confident that this textbook will become an essential resource for educators and students alike, and I look forward to seeing the next generation of intelligent systems developers who will benefit from this work.

University of Macerata   Emanuele Frontoni
Macerata, Italy
VRAI Vision Robotics and Artificial Intelligence Lab
Ancona, Italy

Preface

This book addresses the critical gap between artificial intelligence theory and practical applications. Artificial intelligence is rapidly transforming our world. Understanding both its foundations and implementations has become essential for students and professionals alike.

The chapters follow a carefully designed progression through three main areas. First, we explore the foundational concepts of intelligent systems. Next, we examine optimization and search methods that form the backbone of AI solutions. Finally, we investigate machine learning fundamentals that enable systems to learn from experience and data.

A distinguishing feature of this work is its practical approach. Each theoretical concept is paired with Python implementations and exercises. This hands-on methodology develops both conceptual understanding and practical programming skills simultaneously. The exercises progress from basic implementations to complex real-world problems, allowing readers to build confidence gradually.

The content draws from my experience teaching at Italy's eCampus University and working on industrial applications. I present complex concepts in straightforward language with concrete examples. This approach has proven effective for learners at various levels of technical preparation.

The book serves both undergraduate and graduate students in computer science, engineering, and related disciplines. It assumes basic programming knowledge but introduces concepts progressively to ensure accessibility. Professionals implementing intelligent systems will also find valuable insights and practical guidance throughout the text.

Despite AI's rapid evolution, this book provides both current knowledge and the conceptual framework necessary for understanding future developments. Ethical considerations are addressed throughout, encouraging critical thinking about responsible AI implementation.

I hope this book serves as a valuable resource in your journey to understand and design intelligent systems.

Novedrate, Italy

Oleksandr Kuznetsov

Acknowledgments

With profound gratitude to my Italian friends and the people of Italy, who opened their hearts and homes when we needed it most. Your hospitality and support after our departure from war-torn Ukraine 3 years ago has meant more than words can express.

Special thanks to Emanuele Frontoni and his family, whose friendship and generosity helped us find a new beginning. To all who welcomed us with open arms— your kindness has transformed a difficult journey into a path of hope.

Italy, you have become our second home. This work stands as a testament to the power of human connection across borders and in times of adversity.

Grazie di cuore

Competing Interests The author has no competing interests to declare that are relevant to the content of this manuscript.

About This Book

This book provides a comprehensive introduction to intelligent systems, covering theoretical foundations and practical applications. The content is structured to build knowledge progressively through 18 chapters organized in three thematic parts.

Book Structure

The book is divided into three main parts plus advanced case studies that follow a logical progression:

- Foundations of Intelligent Systems (Chaps. 1–4): These chapters establish the fundamental concepts of artificial intelligence. They explore the historical development of AI, basic principles, and key terminology. This section provides the conceptual framework necessary for understanding more advanced topics.
- Search and Optimization Algorithms (Chaps. 5–14): This section examines various search and optimization methods that are essential for problem-solving in intelligent systems. The chapters cover both classical and modern approaches. Each algorithm is presented with its theoretical basis and practical implementation considerations.
- Machine Learning Fundamentals (Chaps. 15–18): This part focuses on machine learning techniques that enable systems to learn from data. These chapters introduce supervised, unsupervised, and reinforcement learning methods. The section progresses from basic concepts to more sophisticated approaches.

Learning Approach

Each chapter follows a consistent structure designed to enhance understanding:

- Conceptual Introduction: Clear explanations of key ideas with minimal technical jargon.
- Theoretical Framework: Mathematical foundations presented in an accessible manner.
- Practical Implementation: Python code examples demonstrating how to apply concepts.
- Exercises: Problems ranging from basic to advanced to reinforce learning.
- Case Studies: Real-world applications that illustrate practical relevance.

Prerequisites

Readers will benefit most from this book with:

- Basic programming knowledge, particularly in Python
- Foundational understanding of mathematics (algebra, calculus, probability)
- Familiarity with fundamental computer science concepts

No prior knowledge of artificial intelligence or machine learning is required. Necessary concepts are introduced as they appear.

Supporting Materials

All code examples are available online through the companion website. Additional resources include:

- Solution guides for selected exercises
- Supplementary datasets for practice
- Updated examples reflecting current technologies
- Links to relevant research papers

This book serves as both a classroom textbook and a self-study resource for anyone interested in understanding and implementing intelligent systems.

Contents

Abbreviations

ACO	Ant colony optimization
AI	Artificial intelligence
ANN	Artificial neural network
BFGS	Broyden-Fletcher-Goldfarb-Shanno (algorithm)
BFS	Breadth-first search
CNN	Convolutional neural network
DBSCAN	Density-based spatial clustering of applications with noise
DFS	Depth-first search
DL	Deep learning
DNN	Deep neural network
DQN	Deep Q-Network
DRL	Deep reinforcement learning
EA	Evolutionary algorithm
GA	Genetic algorithm
GMM	Gaussian mixture model
HRL	Hierarchical reinforcement learning
IDFS	Iterative deepening depth-first search
KNN	K-nearest neighbors
LSTM	Long short-term memory
MARL	Multi-agent reinforcement learning
MDP	Markov decision process
ML	Machine learning
MSE	Mean squared error
NLP	Natural language processing
PCA	Principal component analysis
POMDP	Partially observable Markov decision process
PSO	Particle swarm optimization
RL	Reinforcement learning
RMSE	Root mean squared error
RNN	Recurrent neural network
SARSA	State-Action-Reward-State-Action (algorithm)

SGD	Stochastic gradient descent
SI	Swarm intelligence
SOM	Self-organizing map
SVM	Support vector machine
TD	Temporal difference
t-SNE	t-Distributed stochastic neighbor embedding
TSP	Traveling salesman problem
UCS	Uniform cost search
VAE	Variational autoencoder

Chapter 1
Introduction to Intelligent Systems

Abstract This chapter presents the fundamental principles of intelligent systems. It explores the defining characteristics of artificial intelligence in contemporary computing environments. The content traces AI's historical development and examines diverse implementations of machine intelligence. Key components that enable computational intelligence are systematically analyzed. The chapter connects theoretical frameworks with practical applications. Through this approach, readers develop both conceptual understanding and implementation skills. This foundation supports further exploration of advanced topics in subsequent chapters. The material balances technical depth with accessibility, making it valuable for both newcomers and those with prior exposure to artificial intelligence concepts.

1.1 Foundations of Intelligence in Computing Systems

The concept of intelligence in computing systems represents one of the most fascinating and rapidly evolving areas in computer science [1, 2]. Understanding what makes a system intelligent provides the foundation for developing advanced artificial intelligence applications [3].

1.1.1 What Makes a System Intelligent?

An intelligent system demonstrates several key characteristics that distinguish it from traditional computing systems. The primary characteristic is adaptability - the ability to modify behavior based on experience and new information [4]. This adaptability manifests through learning capabilities and flexible response to environmental changes.

Intelligence in computing systems requires three fundamental components (Fig. 1.1):

O. Kuznetsov, *Intelligent Systems: From Theory to Applications*, Cognitive Technologies, https://doi.org/10.1007/978-3-032-00044-6_1

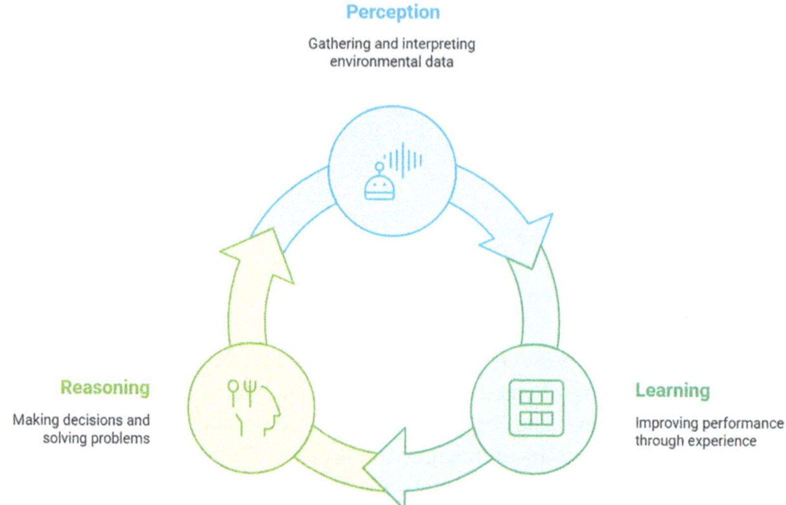

Fig. 1.1 Cycle of Intelligence in Computing Systems

- Perception represents the system's ability to gather and interpret information from its environment. This could range from processing sensor data to analyzing complex datasets.
- Learning enables the system to improve its performance over time. This improvement occurs through experience, pattern recognition, and the ability to generalize from specific examples to broader cases.
- Reasoning allows the system to make decisions and solve problems using available information. This includes both deductive reasoning (drawing conclusions from known facts) and inductive reasoning (forming general principles from specific observations).

These components work together to create what we recognize as intelligent behavior. A chess program that adapts its strategy based on opponent moves demonstrates these principles in action. It perceives the game state, learns from past experiences, and reasons about the best possible moves.

1.1.2 Historical Perspective of Artificial Intelligence

The development of artificial intelligence has evolved through several distinct phases since its inception in the 1950s (Fig. 1.2).

The Early Years (1950–1969) began with the Dartmouth Conference in 1956, which marked the official birth of AI as a field. Researchers focused on symbolic methods and general problem-solving approaches. Allen Newell and Herbert Simon

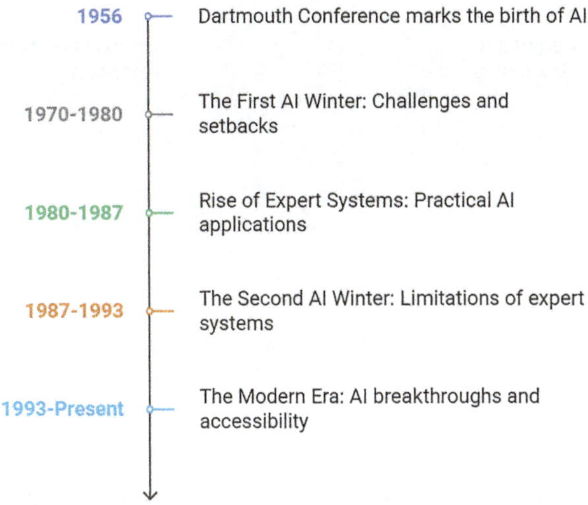

Fig. 1.2 The Evolution of Artificial Intelligence: Key Phases and Milestones

created the Logic Theorist, the first program specifically designed to mimic human problem-solving skills.

The First AI Winter (1970–1980) emerged as early optimism confronted technological limitations. Computing power and memory constraints significantly hindered progress, and many initial predictions proved overly ambitious.

The Expert System Era (1980–1987) demonstrated practical AI applications in specialized domains. XCON, developed for Digital Equipment Corporation, successfully configured computer systems and generated substantial cost savings.

The Second AI Winter (1987–1993) revealed fundamental limitations of expert systems, particularly their inability to learn from experience and adapt to new situations.

The Modern Era (1993–Present) has been characterized by revolutionary developments: increased computing power enabling complex algorithms, vast training resources through big data, breakthroughs in deep learning techniques, and widespread accessibility through cloud computing.

1.1.3   The Turing Test and Fundamental AI Concepts

Alan Turing proposed his famous test in 1950 as a way to evaluate machine intelligence [5, 6]. The test remains influential in AI development and raises fundamental questions about intelligence and consciousness [1, 2].

The basic concept is straightforward [5, 6]: if a human evaluator cannot reliably distinguish between responses from a machine and a human, the machine

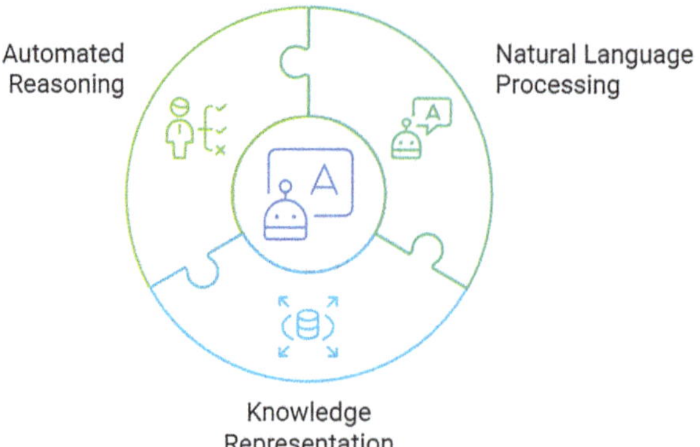

Fig. 1.3 Components of the Turing Test

demonstrates intelligent behavior. The test establishes a practical benchmark for AI systems without requiring a precise definition of intelligence.

Key elements of the Turing Test include (Fig. 1.3):

- Natural Language Processing: The system must communicate effectively in human language. This requires understanding context, nuance, and implicit meaning.
- Knowledge Representation: Information must be stored in a format that enables reasoning and inference. The system needs access to a broad knowledge base about the world.
- Automated Reasoning: The ability to use stored information to answer questions and draw new conclusions proves essential for meaningful conversation.

1.1.4   Types of Intelligence in Computing Systems

Modern AI systems demonstrate various forms of intelligence, each suited to different applications [4]:

1. Narrow AI focuses on specific tasks. These systems excel within their defined domains but cannot transfer knowledge to other areas. Examples include:

   - Chess programs;
   - Image recognition systems;
   - Voice assistants.

2. General AI represents systems that could match human intelligence across all domains. While still theoretical, this remains a major research goal.
   The different types employ various approaches (Fig. 1.4):
   - Rule-Based Systems use predefined rules to make decisions. They excel in well-defined domains with clear rules and boundaries.

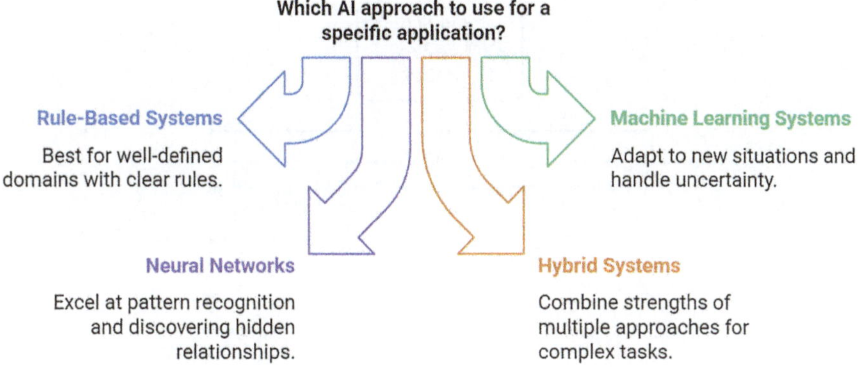

Fig. 1.4 Different types of AI

- Machine Learning Systems learn from experience, improving their performance over time. They adapt to new situations and can handle uncertainty.
- Neural Networks model biological brain structures. They excel at pattern recognition and can discover hidden relationships in data.
- Hybrid Systems combine multiple approaches to leverage their respective strengths. A modern autonomous vehicle might use rule-based systems for basic driving rules while employing neural networks for object recognition.

Each type of intelligence serves specific purposes, and understanding their characteristics helps in selecting appropriate approaches for different applications.

1.2   Core Characteristics of Intelligent Systems

The fundamental characteristics that define intelligent systems emerge from studying both biological intelligence and artificial computational systems [1, 2]. These characteristics enable systems to exhibit intelligent behavior in complex environments.

1.2.1   Learning and Adaptation

Learning represents the most fundamental characteristic of intelligent systems. It enables systems to improve their performance through experience [7, 8]. The learning process involves modifying internal representations and behaviors based on input data and feedback.

Machine learning systems typically employ three main learning paradigms (Fig. 1.5):

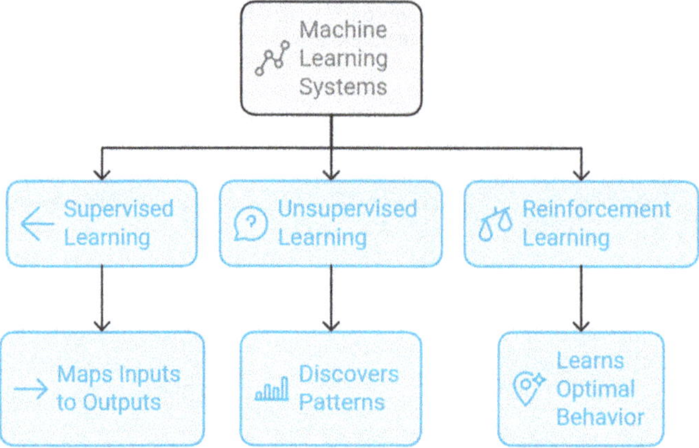

Fig. 1.5 Machine Learning Paradigms

- Supervised Learning involves training with labeled examples. The system learns to map inputs to known outputs. For a given input vector x and output y, the system learns a function $f(x) = y$. This mapping enables prediction on new, unseen data.
- Unsupervised Learning discovers patterns in data without explicit labels. The system identifies inherent structures and relationships. For a dataset X, it finds patterns P such that $P = f(X)$. This enables meaningful data organization and clustering.
- Reinforcement Learning enables systems to learn optimal behavior through interaction with an environment. The system receives rewards or penalties for actions. Over time, it learns a policy $\pi(s)$ that maximizes expected rewards in state s.

1.2.2   Reasoning and Problem Solving

Intelligent systems must process information logically to solve complex problems. This involves several key reasoning mechanisms (Fig. 1.6):

- Deductive Reasoning draws specific conclusions from general principles. Given premises P and a rule $P \Rightarrow Q$, the system concludes Q. This enables precise logical inference.
- Inductive Reasoning generalizes from specific cases to broader principles. From observations $x_1, x_2, \ldots, x_n$ sharing property P, the system infers that similar cases may share P.
- Abductive Reasoning generates the most likely explanation for observations. Given result R and rule $P \Rightarrow R$, the system hypothesizes P as a potential cause.

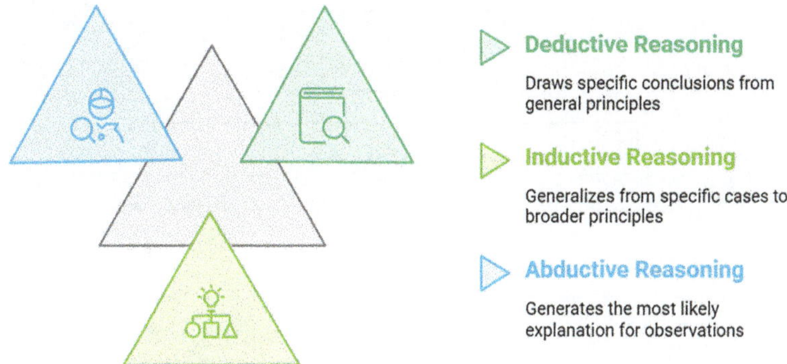

Fig. 1.6 Reasoning Mechanisms in Intelligent Systems

1.2.3   Perception and Environmental Interaction

Perception enables intelligent systems to gather and interpret information from their environment. This process involves several stages [4]:

- Sensor Input Processing converts raw data into usable information. A computer vision system transforms pixel values into meaningful features. The process can be represented as $f(I) = F$, where I represents raw input and F represents extracted features.
- Pattern Recognition identifies meaningful patterns in processed data. The system matches current inputs against known patterns. This enables object recognition, speech understanding, and situation assessment.
- Environmental Mapping creates internal representations of the external world. The system maintains a model M of its environment, updating it as new information arrives:

$$M_{t+1} = update\left(M_{t}, O_{t}\right),$$

- where O_t represents new observations.

1.2.4   Knowledge Representation and Processing

Knowledge representation determines how intelligent systems store and manipulate information. Effective representation enables efficient reasoning and learning (Fig. 1.7).

Symbolic Representation uses explicit symbols and rules. Knowledge is stored in human-readable form. This enables clear reasoning paths but may struggle with uncertainty.

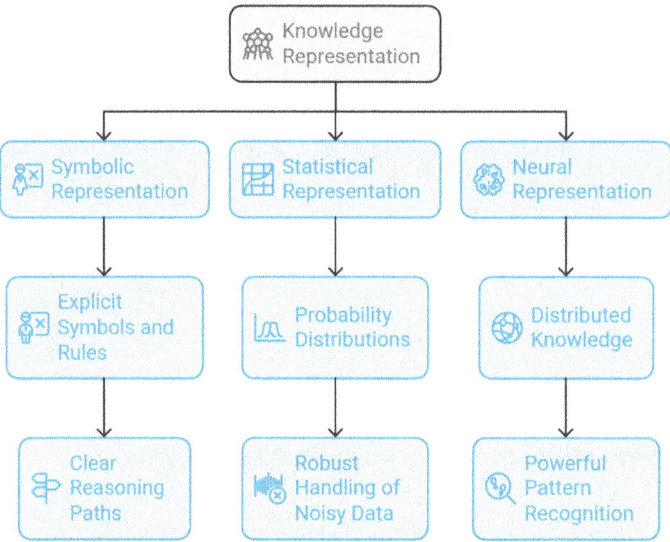

Fig. 1.7 Knowledge Representation in Intelligent Systems

Statistical Representation employs probability distributions and numerical weights. Knowledge includes uncertainty measures. This enables robust handling of noisy data but may lack interpretability.

Neural Representation distributes knowledge across network connections. Information exists in connection weights w_{ij} between nodes. This enables powerful pattern recognition but operates as a "black box."

Knowledge processing involves several operations [1, 2]:

- Retrieval accesses stored information relevant to current tasks. The system must efficiently search its knowledge base.
- Inference generates new knowledge from existing information. This may involve rule application, statistical inference, or pattern completion.
- Update modifies stored knowledge based on new information. The system must maintain consistency while incorporating new data.
- Integration combines information from multiple sources. This requires resolving conflicts and managing uncertainty.

These core characteristics work together to create intelligent behavior. Learning improves reasoning capabilities. Perception provides input for learning. Knowledge representation supports both learning and reasoning. Understanding these relationships helps in designing effective intelligent systems.

Each characteristic can be implemented through various computational approaches. The choice of approach depends on specific application requirements. Simple rule-based systems might suffice for well-defined domains. Complex environments might require sophisticated machine learning and neural networks.

The effectiveness of an intelligent system often depends on how well these characteristics are implemented and integrated. A chess program demonstrates these characteristics working together: it perceives the board state, reasons about possible moves, learns from experience, and maintains knowledge about chess strategies.

1.3   Building Blocks of Intelligent Systems

The construction of intelligent systems requires several essential components working together. These components form a framework that enables intelligent behavior through structured knowledge representation, reasoning capabilities, learning mechanisms, and interaction systems (Fig. 1.8).

1.3.1   Knowledge Base Components

The knowledge base serves as the system's memory, storing facts, rules, and experiences. Its structure significantly impacts system performance and capabilities.

Declarative Knowledge represents facts and relationships. These are statements about the world that can be either true or false. In formal logic notation, we represent a fact as a predicate: $P(x)$ indicates that object x has property P.

Procedural Knowledge contains methods and procedures for using declarative knowledge. This includes algorithms and rules for problem-solving. A rule might take the form: IF $P(x)$ THEN $Q(x)$.

The organization of knowledge often follows hierarchical structures:

- A semantic network connects concepts through defined relationships. Each node represents a concept, and edges represent relationships. This creates a graph structure $G = (V, E)$, where V represents concepts and E represents relationships.

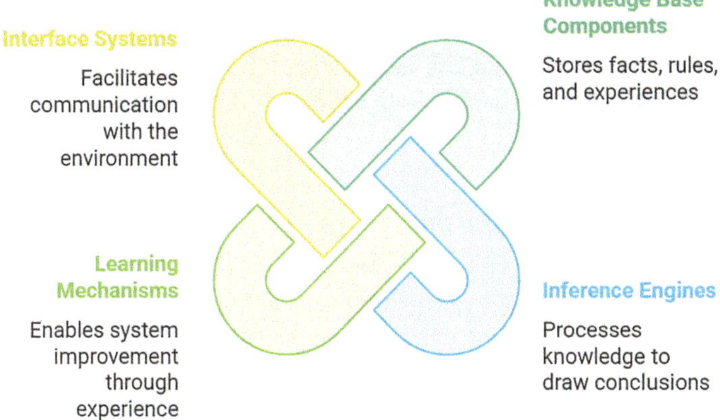

Fig. 1.8 Framework of Intelligent Systems

- Frame-based representation organizes knowledge into structured units. Each frame contains slots for attributes and values. This provides a template-like structure for organizing related information.

1.3.2   Inference Engines

The inference engine processes knowledge to draw conclusions and make decisions [1, 2]. It implements various reasoning mechanisms to generate new knowledge from existing information.

Forward Chaining starts with known facts and applies rules to reach conclusions. The process follows this pattern:

- Match facts against rule conditions;
- Select applicable rules;
- Execute rules to generate new facts;
- Repeat until no new conclusions can be drawn.

Backward Chaining begins with a goal and works backwards to find supporting evidence. The process involves:

- Start with goal state G;
- Find rules with conclusion G;
- Check rule conditions recursively;
- Continue until reaching known facts.

The inference engine must manage uncertainty through mechanisms like [4]:

1. Bayesian Inference updates probabilities based on new evidence:

$$P(H|E) = \frac{P(E|H)P(H)}{P(E)},$$

where H represents a hypothesis and E represents evidence.
2. Fuzzy Logic handles imprecise information through membership functions. A value x belongs to a fuzzy set A with degree $\mu_A(x)$ between 0 and 1.

1.3.3   Learning Mechanisms

Learning mechanisms enable system improvement through experience [7, 8]. These mechanisms modify the knowledge base and decision-making processes based on new information.

Parameter Learning adjusts numerical values in the system. For a neural network, this involves updating weights w_{ij} according to the learning rule:

$$w_{ij}\left(t+1\right) = w_{ij}\left(t\right) + \eta\Delta w_{ij},$$

where η represents the learning rate.

Structure Learning modifies the organization of knowledge. This might involve:

- Creating new rules;
- Forming new connections;
- Pruning ineffective elements.

Memory-Based Learning stores experiences for future reference. The system maintains a case base C containing problem-solution pairs (p_i, s_i). New problems are solved by referring to similar past cases.

1.3.4   Interface and Interaction Systems

Interface systems enable communication between the intelligent system and its environment, including human users and other systems.

Input Processing converts external signals into internal representations. For natural language input, this involves [1, 2]:

- Tokenization;
- Syntactic analysis;
- Semantic interpretation;
- Context integration.

Output Generation produces appropriate responses. This requires:

- Response planning;
- Format selection;
- Content generation;
- Presentation optimization.

Adaptive Interfaces modify their behavior based on user interaction patterns. The interface maintains a user model U that updates with each interaction: $U_{t+1} = update(U_t, I_t)$, where I_t represents the current interaction.

These building blocks must work together seamlessly. The knowledge base provides information for the inference engine. Learning mechanisms update both knowledge and inference rules. Interface systems facilitate information flow between components and the external world.

The effectiveness of an intelligent system depends on proper implementation and integration of these components. Each component must be designed with consideration for [1, 2]:

- Computational efficiency;
- Scalability;
- Reliability;
- Maintainability.

Modern intelligent systems often combine multiple approaches within each component. For example, a knowledge base might use both semantic networks and frames, while the inference engine combines rule-based and probabilistic reasoning.

Understanding these building blocks helps in designing and implementing effective intelligent systems. The choice and configuration of components should align with specific application requirements and constraints.

1.4   Practical Applications and Implementations

Moving from theoretical concepts to practical implementation helps solidify understanding of intelligent systems. This section explores concrete examples and implementations, with a focus on Python-based solutions.

1.4.1   Basic Python Implementation of Intelligent Behaviors

Python offers powerful libraries for implementing intelligent behaviors. Let's explore basic implementations of core intelligent system components.

Simple Rule-Based System:

```python
class RuleBasedSystem:
    def __init__(self):
        self.knowledge_base = {}
        self.rules = []

    def add_fact(self, fact, value):
        self.knowledge_base[fact] = value

    def add_rule(self, condition, conclusion):
        self.rules.append((condition, conclusion))

    def infer(self):
        for condition, conclusion in self.rules:
            if eval(condition, self.knowledge_base):
                self.knowledge_base.update(conclusion)
```

Basic Learning System:

```python
import numpy as np
class SimpleNeuron:
    def __init__(self, learning_rate=0.1):
        self.weights = np.random.rand(2)
        self.learning_rate = learning_rate

    def predict(self, inputs):
        return 1 if np.dot(inputs, self.weights) > 0 else 0

    def train(self, inputs, target):
        prediction = self.predict(inputs)
        error = target - prediction
        self.weights += self.learning_rate * error * inputs
```

1.4.2 Case Studies of Real-World Intelligent Systems

Examining real-world applications provides insight into practical implementation challenges and solutions.

Healthcare Diagnostic System [9, 10]. A modern medical diagnostic system combines multiple AI approaches:

- Knowledge base contains medical conditions and symptoms;
- Neural networks process medical images;
- Rule-based system applies diagnostic guidelines;
- Learning system improves accuracy through experience.

The system might implement diagnostic logic as:

```python
def diagnose(symptoms, vitals, images):
    # Process symptoms through rule-based system
    basic_diagnosis = rule_system.analyze(symptoms)

    # Analyze medical images
    image_findings = neural_net.process(images)

    # Combine evidence for final diagnosis
    final_diagnosis = evidence_combiner(basic_diagnosis,
                                        image_findings,
                                        vitals)
    return final_diagnosis
```

Autonomous Navigation System [11, 12]. Modern autonomous vehicles demonstrate complex intelligent system integration:

- Sensor processing for environmental perception;
- Real-time decision making for navigation;
- Learning from driving experiences;
- Safety rule enforcement.

1.4.3   Ethical Considerations in System Design

Ethical implementation of intelligent systems requires careful consideration of several key aspects [1, 2].

1. Fairness and Bias. Systems must be tested for bias across different groups. A simple bias check might look like:

```
def check_bias(predictions, demographic_groups):
    for group in demographic_groups:
        accuracy = compute_accuracy(predictions[group])
        if abs(accuracy - baseline_accuracy) > threshold:
            raise BiasWarning(f"Potential bias for {group}")
```

2. Privacy Protection. Data handling should follow strict privacy guidelines:

```
class PrivateData:
    def __init__(self, data):
        self.data = self.anonymize(data)

    def process(self, operation):
        if self.check_privacy_compliance(operation):
            return operation(self.data)
            raise PrivacyViolation
```

1.4.4   Future Trends and Challenges

The field of intelligent systems continues to evolve rapidly [1, 2]. Several key trends and challenges shape its development.

1. Enhanced Learning Capabilities. Future systems will likely feature:

- More efficient learning from limited data;

- Better transfer learning between domains;
- Improved unsupervised learning capabilities.

The basic framework for advanced learning might look like:

```python
class AdvancedLearner:
    def learn_from_few_examples(self, examples):
        base_knowledge = self.extract_principles(examples)
        return self.generalize(base_knowledge)

    def transfer_knowledge(self, source_domain, target_domain):
        common_features = self.find_similarities(
            source_domain, target_domain)
        return self.adapt_knowledge(common_features)
```

2. Resource Efficiency. Systems must become more efficient in terms of:

- Computational requirements;
- Energy consumption;
- Memory usage.

A framework for resource monitoring might include:

```python
class ResourceManager:
    def monitor_usage(self):
        cpu_usage = self.measure_cpu()
        memory_usage = self.measure_memory()
        energy_usage = self.measure_energy()

        if any([cpu_usage > cpu_threshold,
                memory_usage > memory_threshold,
                energy_usage > energy_threshold]):
            self.optimize_resources()
```

3. Integration Challenges. Future systems will need to:

- Combine multiple AI approaches seamlessly;
- Interface with legacy systems;
- Scale efficiently with growing data volumes;
- Maintain reliability under varying conditions.

The implementation of intelligent systems requires balancing theoretical understanding with practical constraints. Success depends on careful consideration of:

- System requirements;
- Available resources;

- Ethical implications;
- Future maintainability;
- User needs.

Each implementation should be viewed as an iterative process, with continuous evaluation and improvement based on real-world performance and feedback.

1.5   Practice and Exercises

This section provides hands-on exercises to reinforce understanding of intelligent systems through practical implementation and experimentation. All exercises are designed to build both conceptual understanding and practical skills.

1.5.1   Self-Assessment Questions

These questions are designed to test your understanding of key concepts presented in the chapter. For each question, select the answer you believe is correct. This exercise will help solidify your knowledge and identify areas that may need additional review.

1. What primarily distinguishes an intelligent system from a traditional system?

    (a) Processing speed
    (b) Ability to adapt to new situations
    (c) Amount of data processed
    (d) Advanced user interface

2. The concept of "learning" in an intelligent system refers to:

    (a) Acquiring new data from an external database
    (b) Ability to improve performance based on experience
    (c) Manual software updates by developers
    (d) Storing large amounts of information

3. Which component of an intelligent system is responsible for decision-making?

    (a) Knowledge base
    (b) User interface
    (c) Inference engine
    (d) Input/output module

4. The "connectionist" approach to artificial intelligence primarily relies on:

    (a) Predefined logical rules
    (b) Artificial neural networks

(c) Advanced search algorithms
(d) Expert systems

5. When was the "AI winter" period?

(a) 1950s and 1960s
(b) 1970s and 1980s
(c) 1990s
(d) 2000s

1.5.2 Critical Thinking Exercise

Objective: Apply concepts learned about intelligent systems to a real-world case study.

Task: Consider a modern voice assistant (like Siri, Alexa, or Google Assistant).

A. Identify and describe at least three characteristics of intelligent systems present in these assistants.
B. Choose one of the following ethical or technical challenges these systems might face in the future and discuss it briefly (approximately 100 words):

- User data privacy and security;
- Bias in language interpretation or responses;
- User dependency on voice assistants;
- Impact on job market (e.g., replacement of human workers).

C. Propose a possible solution or approach to address the challenge you discussed.

Format: Structure your response in three distinct paragraphs, corresponding to points A, B, and C. Total length should not exceed 300 words.

1.6 Conclusions and Future Directions

This chapter has established the fundamental concepts of intelligent systems and their practical applications. The integration of learning, reasoning, and adaptation capabilities creates systems that can address increasingly complex challenges across diverse domains from healthcare to autonomous navigation.

The development of intelligent systems requires balancing technological innovation with ethical responsibility. Several trends will shape the future of this field:

- Advanced learning mechanisms that enable more efficient knowledge acquisition and transfer
- Seamless integration of multiple AI approaches to create more robust systems

– Resource-efficient implementations that reduce computational and energy requirements

Key challenges ahead include developing more robust learning algorithms, ensuring fairness and preventing bias, protecting privacy, maintaining transparency, and effectively integrating intelligent systems with existing infrastructure.

The principles, tools, and considerations discussed in this chapter provide a foundation for more advanced study and practical implementation of intelligent systems in subsequent chapters.

References

1. Russell, S., Norvig, P.: Artificial Intelligence: A Modern Approach. Pearson, Hoboken (2020).
2. Norvig, P.: Artificial intelligence: a modern approach. Global edition. Pearson, Boston (2021).
3. Géron, A.: Hands-On Machine Learning with Scikit-Learn, Keras, and TensorFlow: Concepts, Tools, and Techniques to Build Intelligent Systems. O'Reilly Media, Beijing China; Sebastopol, CA (2019).
4. Artificial Intelligence: Principles and Practice.
5. TURING, A.M.: I.—COMPUTING MACHINERY AND INTELLIGENCE. Mind. LIX, 433–460 (1950). https://doi.org/10.1093/mind/LIX.236.433.
6. Oppy, G., Dowe, D.: The Turing Test. (2003).
7. Goodfellow, I., Bengio, Y., Courville, A.: Deep Learning. The MIT Press, Cambridge, Massachusetts (2016).
8. Bishop, C.M., Bishop, H.: Deep Learning: Foundations and Concepts. Springer, Cham (2023).
9. Vannaprathip, N., Haddawy, P., Schultheis, H., Suebnukarn, S.: SDMentor: A virtual reality-based intelligent tutoring system for surgical decision making in dentistry. Artificial Intelligence in Medicine. 103092 (2025). https://doi.org/10.1016/j.artmed.2025.103092.
10. Fu, L., Wu, X., Lou, X., Zhang, Q., Qiu, D.: Intelligent Analgesia Management System in Postoperative Pain Management: A Retrospective Analysis. Journal of PeriAnesthesia Nursing. (2025). https://doi.org/10.1016/j.jopan.2024.08.015.
11. Guo, L., Li, W., Liu, X., Hu, P., Tian, B., Yang, J.: BIO-Inspired Intelligent Navigation. In: Reference Module in Materials Science and Materials Engineering. Elsevier (2024). https://doi.org/10.1016/B978-0-443-14081-5.00011-8.
12. Zhang, G., Hsu, L.-T.: Intelligent GNSS/INS integrated navigation system for a commercial UAV flight control system. Aerospace Science and Technology. 80, 368–380 (2018). https://doi.org/10.1016/j.ast.2018.07.026.

Chapter 2
The Evolution of Artificial Intelligence

Abstract This chapter traces artificial intelligence's evolution from theoretical concepts to practical implementations. It examines pivotal milestones, influential pioneers, and technological breakthroughs that have defined the field. The narrative moves chronologically through distinct developmental phases, revealing both spectacular advances and significant setbacks. Special attention is given to the cyclical nature of AI progress, particularly during periods of diminished expectations ("AI winters") and subsequent resurgences.

2.1 Foundations of AI (1940–1956)

The foundations of artificial intelligence emerged from the convergence of mathematics, psychology, and early computer science in the mid-twentieth century. This period established fundamental concepts that continue to shape modern AI development.

2.1.1 Early Computing and Logic Concepts

The concept of intelligent machines first gained serious consideration with the development of electronic computers in the 1940s [1]. Early pioneers recognized that computers could potentially process more than just numerical calculations. They envisioned machines that could manipulate symbols and engage in logical reasoning.

The development of Boolean logic provided a crucial mathematical foundation [2]. Boolean algebra reduced logical operations to simple true/false statements. This binary approach aligned perfectly with the fundamental operation of digital computers. Engineers could now represent complex logical relationships through combinations of simple binary operations.

O. Kuznetsov, *Intelligent Systems: From Theory to Applications*, Cognitive
Technologies, https://doi.org/10.1007/978-3-032-00044-6_2

Claude Shannon's works on information theory in 1948 provided another essential foundation [3, 4]. Shannon demonstrated how information could be quantified and processed mathematically. His theories showed how machines could handle not just numbers, but meaningful information. This breakthrough opened new possibilities for machine intelligence.

2.1.2   *The Turing Test and Its Implications*

Alan Turing's 1950 paper "Computing Machinery and Intelligence" marked a watershed moment in AI history [5]. Instead of asking the philosophical question "Can machines think?", Turing proposed a practical test of machine intelligence. The test, now known as the Turing Test, offered a concrete way to evaluate machine intelligence.

The test procedure is remarkably straightforward [6]. A human judge engages in text conversations with both a human and a machine. If the judge cannot reliably distinguish between the human and machine responses, the machine passes the test. This simple framework provided the first operational definition of machine intelligence.

Turing's insights went beyond the test itself. He addressed key objections to the possibility of machine intelligence. These included concerns about consciousness, creativity, and emotional capacity. His responses to these objections remain relevant to modern AI discussions.

2.1.3   *McCulloch-Pitts Neural Networks*

In 1943, Warren McCulloch and Walter Pitts introduced the first mathematical model of an artificial neuron [7]. Their model showed how simple computing units could represent logical operations. A McCulloch-Pitts neuron accepts multiple binary inputs and produces a binary output based on a threshold function [7]:

$$\text{Output} = \begin{cases} 1 & \text{if } \sum_{i=1}^{n} w_i x_i \geq \theta; \\ 0 & \text{otherwise,} \end{cases}$$

where x_i represents inputs, w_i represents weights, and θ is the threshold.

This simple model demonstrated several key principles (Fig. 2.1):

- Neural networks could implement logical functions;
- Simple units could combine to perform complex computations;
- Binary thresholds could model biological neuron behavior.

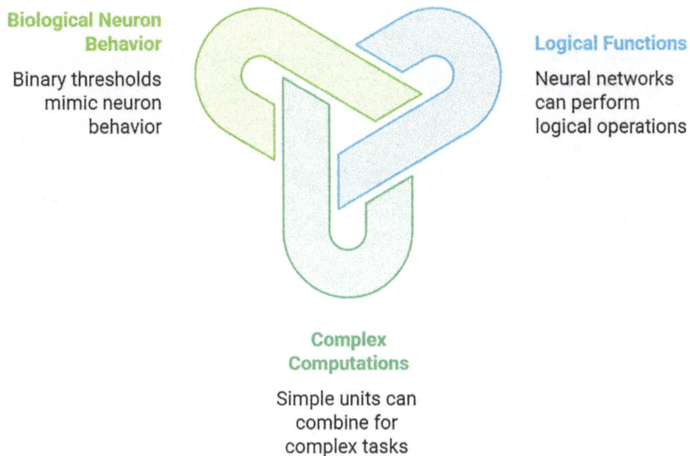

Fig. 2.1 Key Principles of McCulloch-Pitts Neural Networks

2.1.4   The Dartmouth Conference and Birth of AI

The 1956 Dartmouth Summer Research Project marked the official birth of artificial intelligence as a field. John McCarthy, Marvin Minsky, Nathaniel Rochester, and Claude Shannon organized this pivotal eight-week workshop [10, 11]. The conference proposal contained the first use of the term "artificial intelligence."

The conference gathered leading researchers to explore ways machines could simulate human intelligence. Key topics included:

- Natural language processing;
- Neural networks;
- Theory of computation;
- Abstract reasoning;
- Creativity.

While the conference did not achieve its ambitious goals, it established AI as a distinct academic discipline [10, 11]. It created a community of researchers dedicated to the development of intelligent machines. The relationships formed at Dartmouth shaped AI research for decades to come.

The participants shared a fundamental optimism about AI's potential. They believed that significant advances were possible within a generation. Though this timeline proved overly optimistic, their core insights about machine intelligence remain relevant today [8, 9].

This foundational period established key principles that still guide AI development (Fig. 2.2):

- The importance of symbolic reasoning;
- The potential of neural computation;

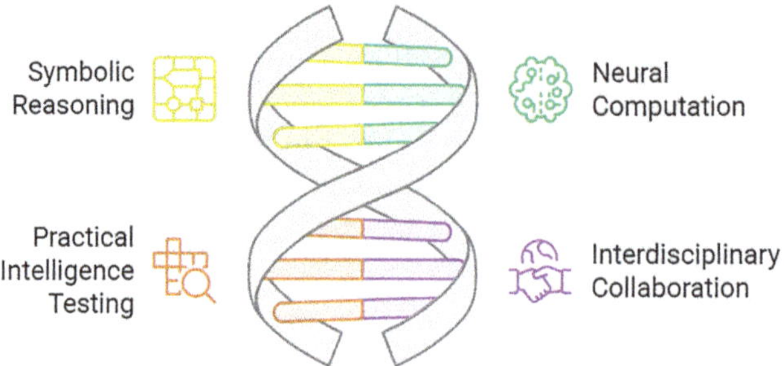

Fig. 2.2 Key Principles Shaping the Future of AI Development

- The need for practical tests of intelligence;
- The value of interdisciplinary collaboration.

Understanding these foundations helps contextualize modern AI developments. Many current challenges and approaches have roots in this formative period.

2.2 The First Golden Age (1956–1974)

The period following the Dartmouth Conference marked the first golden age of artificial intelligence [10, 11]. This era saw remarkable optimism and significant technical achievements. Researchers made groundbreaking progress in several key areas of AI development.

2.2.1 Early AI Programs and Achievements

The Logic Theorist, developed by Allen Newell and Herbert Simon in 1956, represented the first AI program specifically designed for problem-solving [12]. The program could prove mathematical theorems by mimicking human problem-solving strategies. It successfully proved 38 of the first 52 theorems from Principia Mathematica, demonstrating that machines could perform tasks requiring intelligence [8, 9].

The General Problem Solver (GPS), created in 1957, took problem-solving abilities further [13, 14]. GPS could break complex problems into simpler sub-goals. This approach, called means-ends analysis, remains influential in modern AI. The system demonstrated how computers could approach problems in a way similar to human thinking.

2.2.2   Development of Natural Language Processing

Natural language processing emerged as a crucial focus of early AI research. Joseph Weizenbaum's ELIZA program, developed in 1966, marked a significant milestone [15]. ELIZA simulated a psychotherapist by recognizing patterns in user input and generating appropriate responses. While simple by modern standards, ELIZA demonstrated the potential for human-computer interaction through natural language.

SHRDLU, created by Terry Winograd in 1970, represented a more sophisticated approach to language understanding [16]. The program could interpret and respond to natural language commands about a simple block world. SHRDLU demonstrated comprehension of context and basic reasoning about physical objects. The system could maintain a coherent dialogue about its actions and the state of its world.

2.2.3   Pattern Recognition Advances

Early pattern recognition systems focused on visual processing and character recognition. The first successful optical character recognition (OCR) systems emerged during this period. These systems could convert printed text into machine-readable format, though with limited accuracy.

Frank Rosenblatt's Perceptron, introduced in 1957 [17, 18], provided a mathematical model for pattern recognition. The Perceptron learning rule can be expressed as:

$$w_i(t+1) = w_i(t) + \eta(d-y)x_i,$$

where w_i represents weights, η is the learning rate, d is the desired output, y is the actual output, and x_i represents inputs.

2.2.4   Early Expert Systems

The first expert systems emerged as attempts to capture human expertise in specific domains. DENDRAL, developed at Stanford in 1965, became the first successful expert system [14, 19]. It could analyze mass spectrometry data to identify chemical compounds. The system demonstrated that computers could perform specialized analytical tasks at an expert level.

These early expert systems established key principles (Fig. 2.3):

- Knowledge representation through rules and facts;
- Inference mechanisms for decision-making;
- Explanation capabilities for system conclusions;
- Domain-specific expertise encoding.

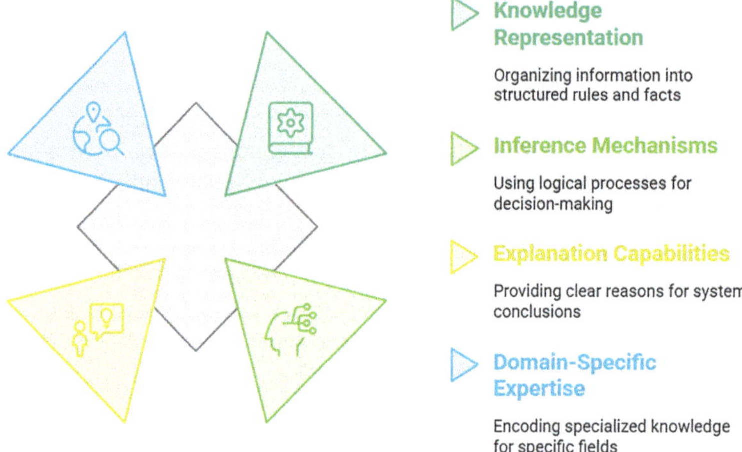

Fig. 2.3 Key Principles of Expert Systems

The success of these early systems generated significant optimism about AI's potential. Researchers and funding agencies believed that human-level artificial intelligence was within reach. Many predicted major breakthroughs within a decade.

Characteristic features of this golden age included [14]:

- Focus on general problem-solving methods;
- Development of foundational algorithms;
- Emphasis on symbolic reasoning;
- Optimistic predictions about AI's future.

This period established many core concepts that remain relevant today. Modern AI systems still build upon these early achievements. Understanding this era helps contextualize current developments in artificial intelligence.

However, the optimism of this period would soon face significant challenges. The limitations of early approaches would become increasingly apparent. These challenges would eventually lead to the first "AI winter." Yet the foundational work of this golden age created the basis for all subsequent AI development.

This era demonstrated both the potential and limitations of early AI approaches. It showed that machines could perform intelligent tasks in specific domains. However, it also revealed the complexity of creating general artificial intelligence.

2.3 The First AI Winter and Recovery (1974–1980)

The optimism of AI's first golden age gave way to a period of reduced funding and diminished expectations. This period, known as the first "AI winter," emerged from growing recognition of fundamental limitations in early AI approaches.

2.3.1   Lighthill Report and Its Impact

The Lighthill Report, commissioned by the British government in 1973, delivered a critical assessment of artificial intelligence research [20]. Sir James Lighthill, a prominent mathematician, examined the field's progress and prospects. His conclusions significantly influenced AI funding and research directions.

The report identified several key problems [14, 20]:

- Limited progress compared to early predictions;
- Inability of AI systems to scale beyond simple problems;
- Lack of practical applications despite significant investment;
- Fundamental limitations in available computing power.

Following the report's publication, research funding for AI projects decreased dramatically in the UK. This reduction in support spread to other countries, particularly affecting basic research in artificial intelligence. Many researchers shifted their focus to more specialized and immediately practical applications.

2.3.2   Limitations of Early Approaches

The AI winter revealed fundamental limitations in early approaches to artificial intelligence [14, 20]. The "combinatorial explosion" problem became increasingly apparent. As problems grew more complex, the computational requirements grew exponentially. Simple algorithms that worked well for toy problems failed when applied to real-world situations.

Early natural language processing systems demonstrated severe limitations. Programs like ELIZA showed the difficulty of achieving true language understanding. The systems could not handle context, ambiguity, or common-sense reasoning. These limitations highlighted the gap between simulation and genuine comprehension.

Knowledge representation proved more challenging than expected. Early systems struggled to capture and use complex real-world knowledge. The difficulty of encoding common-sense knowledge became a major obstacle. This challenge, known as the "common sense problem," remains relevant today.

2.3.3   Emergence of New Paradigms

The setbacks of the AI winter prompted researchers to explore new approaches. The field began shifting from general problem-solving methods to more specialized techniques. This shift emphasized practical applications over general artificial intelligence [10, 14].

Researchers developed more sophisticated methods for handling uncertainty. Probabilistic approaches gained prominence. These methods could better handle the ambiguity inherent in real-world problems. The integration of statistical techniques with traditional AI methods began during this period.

The importance of domain knowledge became increasingly recognized. Researchers focused on building systems with deep knowledge of specific domains. This approach led to more practical and successful applications, particularly in expert systems.

2.3.4 Lessons Learned from Early Setbacks

The AI winter provided valuable lessons for the field's development:

- First, the danger of overpromising became clear. Early predictions about AI's capabilities had been too optimistic. The field learned the importance of realistic goal-setting and careful evaluation of progress.
- Second, the value of practical applications emerged. Successful AI systems needed to solve real problems rather than just demonstrate interesting capabilities. This understanding led to greater emphasis on applicable research.
- Third, the limitations of pure symbolic approaches became evident. The field began recognizing the need for multiple approaches to intelligence. This recognition led to increased interest in neural networks and statistical methods.
- Fourth, the importance of computational efficiency gained recognition. Successful AI systems needed to operate within real-world computational constraints. This understanding influenced algorithm development and system design.

These lessons shaped the recovery period that followed the AI winter [10, 14]:

- Greater emphasis on practical applications;
- More realistic assessment of capabilities;
- Increased focus on specialized systems;
- Better integration of multiple approaches.

The first AI winter, while challenging for the field, ultimately contributed to its maturation. It prompted more rigorous evaluation of methods and more practical approaches to development. These changes laid the groundwork for subsequent advances in artificial intelligence.

Understanding this period helps modern AI practitioners avoid similar pitfalls. It reminds us of the importance of balancing ambition with practical constraints. The lessons learned continue to influence how we approach artificial intelligence development today.

2.4   The Expert Systems Era (1980–1987)

The period from 1980 to 1987 marked a significant revival in artificial intelligence through the development and commercial success of expert systems [10, 14]. This era demonstrated the practical value of AI in solving real-world problems.

2.4.1   Knowledge-Based Systems

Expert systems represented a new approach to artificial intelligence [14, 19]. These systems captured human expertise in specific domains through explicit rules and facts. The knowledge base contained domain-specific information structured for computer reasoning.

A typical expert system consisted of three main components:

- Knowledge Base: Stored facts and rules about the domain;
- Inference Engine: Applied the rules to solve problems;
- User Interface: Facilitated interaction with users.

The rules typically followed an IF-THEN structure:

```
IF (condition1 AND condition2)
THEN (conclusion OR action)
For example, a medical diagnosis system might use rules like:
Copy
IF (temperature > 38.5°C AND sore_throat = true)
THEN (suggest_diagnosis = "possible strep infection")
```

2.4.2   Commercial Applications

MYCIN, developed at Stanford between 1972 and 1980, represented an early commercial success [19, 21]. The system diagnosed blood infections and recommended antibiotics. MYCIN demonstrated performance comparable to human experts in its specific domain.

DENDRAL became the first expert system used outside academic settings [22]. It assisted chemists in identifying molecular structures from mass spectrometry data. The system proved particularly valuable in analyzing complex organic compounds.

2.4.3   XCON and Other Success Stories

XCON (originally called R1) marked a major commercial triumph for expert systems [23]. Developed for Digital Equipment Corporation, XCON configured computer systems. The system automated a complex task that previously required human experts.

XCON's success demonstrated several key benefits:

- Reduced configuration errors;
- Faster processing of orders;
- Significant cost savings;
- Consistent decision-making.

Digital Equipment Corporation reported saving millions of dollars annually through XCON's use. This success prompted other companies to invest in expert system development.

2.4.4   Limitations and Challenges

Despite their successes, expert systems faced significant limitations [19, 21]. Knowledge acquisition proved particularly challenging. Converting human expertise into explicit rules required extensive time and effort. This process, known as the "knowledge acquisition bottleneck," limited system development.

Maintenance of expert systems presented another major challenge. Rules often interacted in complex ways. Making changes to one part of the system could have unexpected effects elsewhere. Large rule bases became increasingly difficult to maintain and update.

Expert systems struggled with [19, 21]:

- Handling uncertainty and incomplete information;
- Adapting to new situations;
- Learning from experience;
- Dealing with exceptional cases.

These systems operated effectively only within their specific domains. They could not generalize knowledge or adapt to new situations. This brittleness limited their broader applicability.

2.4.5   The Rise and Decline

The success of early expert systems led to increased investment in AI [8]. Companies created specialized hardware and software for expert system development. New companies formed to develop and market expert system technology.

However, limitations became increasingly apparent:

- High development costs;
- Difficulty in maintaining large systems;
- Inability to learn or adapt;
- Limited scope of application.

By the late 1980s, the limitations of expert systems led to decreased enthusiasm. The specialized hardware became obsolete as general-purpose computers grew more powerful. Many companies that had invested heavily in expert system technology faced difficulties.

2.4.6   Lasting Impact

Despite their limitations, expert systems provided valuable lessons for AI development (Fig. 2.4):

- The importance of domain-specific knowledge;
- The value of explainable decision-making;
- The need for practical, focused applications;
- The challenges of knowledge representation.

Modern AI systems still incorporate many insights from the expert systems era. The emphasis on capturing and using domain knowledge remains relevant. The experience with expert systems continues to influence how we approach artificial intelligence development.

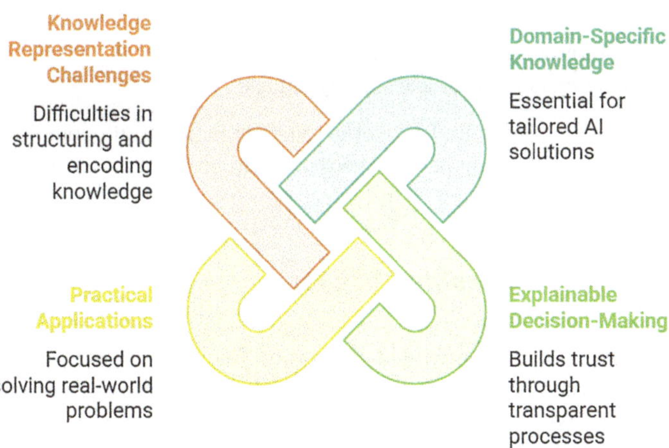

Fig. 2.4 Lessons from Expert Systems

This era demonstrated both the potential and limitations of rule-based approaches to AI. It showed that artificial intelligence could solve practical problems while highlighting the challenges of creating truly intelligent systems.

2.5 Machine Learning Revolution (1990–2000)

The 1990s marked a fundamental shift in artificial intelligence [10, 14]. The field moved from rule-based systems toward methods that could learn from data. This transformation revolutionized how we approach artificial intelligence.

2.5.1 Neural Networks Renaissance

Neural networks experienced a significant revival in the 1990s [10, 14]. The back-propagation algorithm, though discovered earlier, became practical with increased computing power [24, 25]. This algorithm allowed neural networks to learn from examples by adjusting their internal weights.

The basic principle of backpropagation can be expressed as [24, 25]:

$$w_{ij}\left(t+1\right) = w_{ij}\left(t\right) - \eta \frac{\partial E}{\partial w_{ij}},$$

where w_{ij} represents the connection weights, η is the learning rate, and E is the error function.

Researchers made several key improvements to neural network technology [8, 9]:

- Better initialization methods;
- More effective training algorithms;
- New network architectures;
- Improved understanding of learning dynamics.

These advances enabled neural networks to tackle increasingly complex problems. Applications expanded from pattern recognition to prediction and control tasks.

2.5.2 Statistical Learning Methods

Statistical learning theory provided a rigorous foundation for machine learning. This approach combined probability theory with computational learning methods. It helped explain why learning algorithms work and when they might fail.

Key concepts from statistical learning included [8, 9]:

- Bias-variance tradeoff;
- Overfitting and underfitting;
- Generalization bounds;
- Model complexity measures.

These theoretical insights led to practical improvements in learning algorithms. They helped researchers design more effective and reliable learning systems.

2.5.3   Support Vector Machines

Support Vector Machines (SVM), developed by Vladimir Vapnik and colleagues, represented a major breakthrough [26]. SVMs provided a new approach to classification and regression problems [27, 28]. The method found optimal boundaries between different classes of data.

The basic SVM optimization problem aims to maximize the margin between classes [27, 28]:

$$maximize \ \frac{2}{|w|} \ \text{subject to} \ y_i \left(w \cdot x_i + b \right) \geq 1,$$

where w is the normal vector to the separating hyperplane, and b is the bias term.
SVMs offered several advantages:

- Strong theoretical foundations;
- Effective handling of high-dimensional data;
- Robust performance;
- Ability to capture non-linear patterns.

These characteristics made SVMs particularly valuable for practical applications like text classification and bioinformatics.

2.5.4   Practical Applications

The machine learning revolution enabled new practical applications [9, 10]. Speech recognition systems became more accurate and practical. Computer vision systems could better recognize objects and faces.

Machine learning found applications in (Fig. 2.5):

- Financial forecasting;
- Medical diagnosis;
- Industrial control;

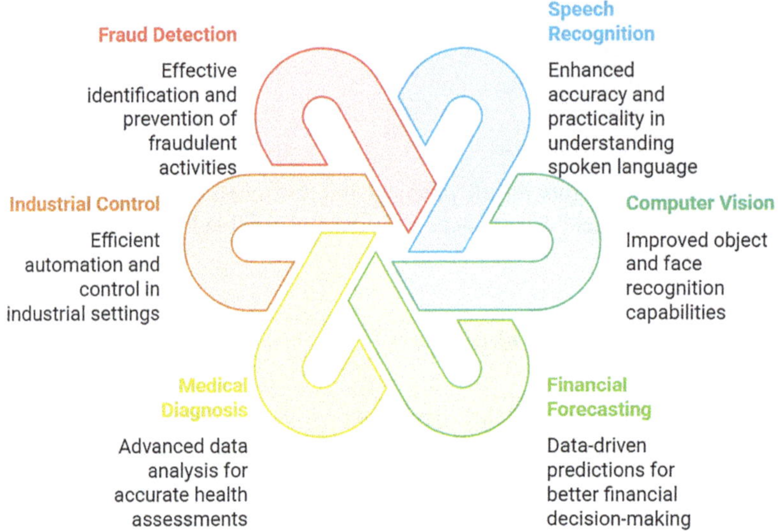

Fig. 2.5 The Impact of Machine Learning

- Fraud detection;
- Recommendation systems.

Each application demonstrated the power of learning from data rather than relying on hand-coded rules.

2.5.5   The Shift in Approach

This period marked a fundamental change in how we approach AI problems. Instead of trying to program intelligence directly, systems learned from examples. This data-driven approach proved more effective for many real-world problems.

The success of machine learning methods led to (Fig. 2.6):

- Increased focus on data collection;
- Development of better learning algorithms;
- New tools for data analysis;
- Growing commercial applications.

Key innovations during this period included:

- Ensemble methods like Random Forests;
- Boosting algorithms;
- New clustering techniques;
- Improved optimization methods.

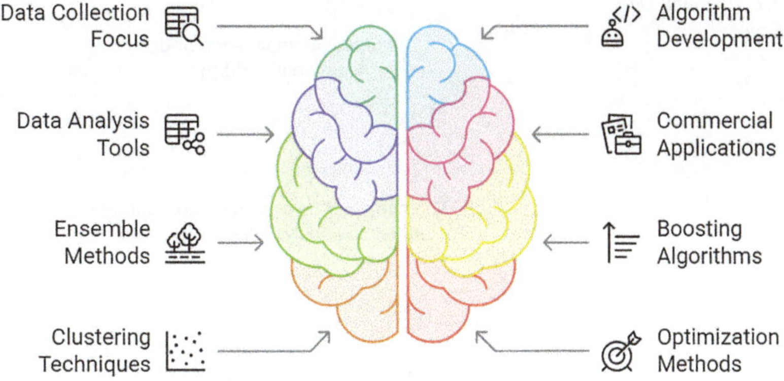

Fig. 2.6 Evolution of Machine Learning

These developments laid the groundwork for future advances in deep learning and artificial intelligence. They established machine learning as a central paradigm in AI research and application.

The machine learning revolution demonstrated the value of learning from data. It showed that many intelligent behaviors could emerge from statistical patterns in data. This insight continues to guide modern artificial intelligence development.

Understanding this period helps explain current trends in AI. Many modern developments build directly on the foundations established during the machine learning revolution. The principles developed during this time remain fundamental to how we approach artificial intelligence today.

2.6   The Deep Learning Era (2000–Present)

The early twenty-first century witnessed an unprecedented transformation in artificial intelligence through deep learning [10, 14]. This approach revolutionized the field and enabled capabilities previously considered impossible.

2.6.1   Big Data and Computing Power

Two critical developments enabled the deep learning revolution [29, 30]. First, the exponential growth in digital data provided the raw material for training sophisticated models. The internet, mobile devices, and digital sensors generated vast quantities of text, images, video, and other data types (Fig. 2.7).

Second, dramatic increases in computing power made complex neural network training practical. Graphics Processing Units (GPUs), originally designed

Fig. 2.7 Enabling Deep
Learning

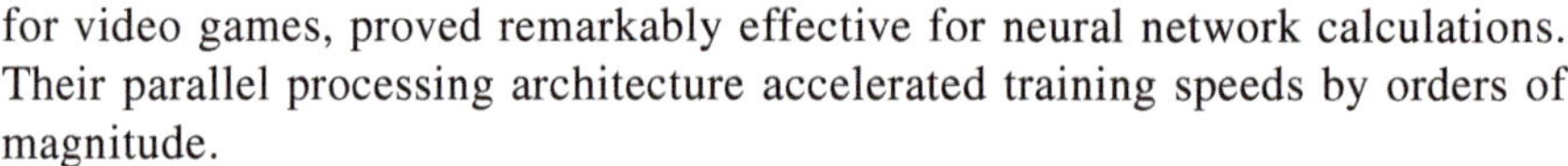

for video games, proved remarkably effective for neural network calculations. Their parallel processing architecture accelerated training speeds by orders of magnitude.

Cloud computing further transformed AI development by making powerful computational resources widely accessible. Researchers and companies could now access thousands of processors without massive infrastructure investments.

The combination of data and computing power created the perfect environment for deep learning. Algorithms that had existed for decades suddenly became practical. Neural networks with many layers—once considered too difficult to train—now demonstrated remarkable capabilities.

2.6.2   *Breakthrough Applications*

Deep learning produced several landmark achievements that captured public attention [29, 30]. In 2012, AlexNet dramatically improved image recognition accuracy in the ImageNet competition [31]. This breakthrough demonstrated deep learning's potential for visual tasks.

IBM's Watson system [32] defeated human champions in Jeopardy! in 2011. While not purely a deep learning system, Watson showcased the potential of AI in natural language processing and question answering.

DeepMind's AlphaGo [33] defeated the world champion Go player in 2016. This achievement came decades earlier than experts had predicted. It demonstrated that AI could master tasks requiring intuition and strategic thinking previously considered uniquely human.

Deep learning transformed numerous fields (Fig. 2.8):

- Computer vision systems now recognize objects, faces, and activities;
- Speech recognition reached human-level accuracy in many contexts;

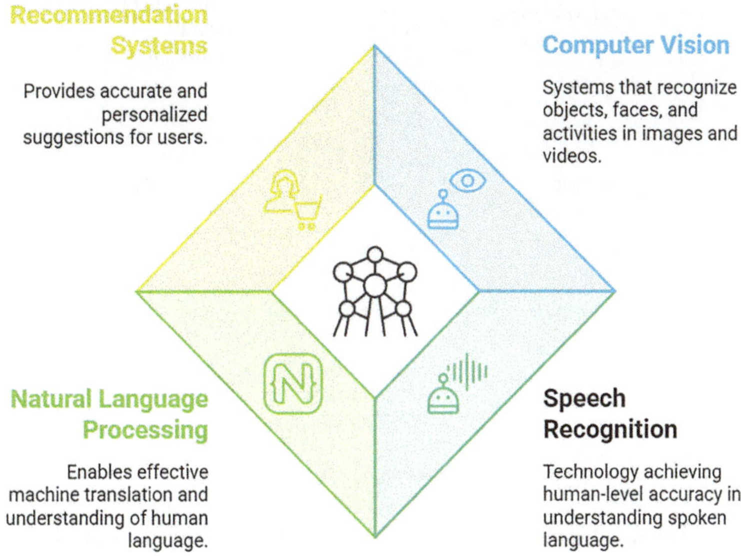

Fig. 2.8 Transformative Impact of Deep Learning on Everyday Technology

- Natural language processing enabled effective machine translation;
- Recommendation systems became more accurate and personalized.

These advances moved AI from research labs into everyday applications. Smartphones, digital assistants, and online services now routinely use deep learning technology.

2.6.3   Key Technological Advances (Fig. 2.9)

Convolutional Neural Networks (CNNs) revolutionized computer vision [31, 34]. These networks use specialized layers that mimic aspects of the human visual system. A typical convolutional layer applies filters across an image [34]:

$$a_{i,j}^{(l)} = \sigma\left(\sum_{m=0}^{F-1}\sum_{n=0}^{F-1} w_{m,n}^{(l)} a_{i+m,j+n}^{(l-1)} + b^{(l)} \right),$$

where $a_{i,j}^{(l)}$ represents the activation at position (i,j) in layer l, $w_{m,n}^{(l)}$ represents filter weights, and σ is an activation function.

Recurrent Neural Networks (RNNs) and Long Short-Term Memory (LSTM) networks enabled processing of sequential data [35]. These architectures maintain internal state, allowing them to process sequences like text, speech, and time series data effectively.

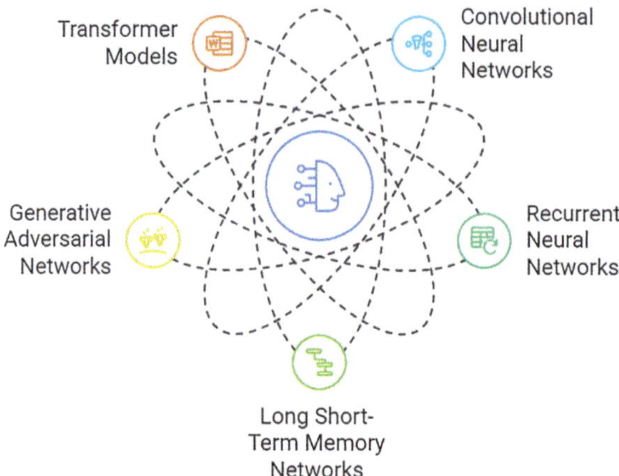

Fig. 2.9 Advances in AI and Machine Learning

Generative Adversarial Networks (GANs) [36], introduced in 2014, created a new approach to generating content. GANs consist of two networks—a generator and a discriminator—competing against each other. This competition produces increasingly realistic synthetic content including images, music, and text.

Transformer models [8, 37], introduced in 2017, revolutionized natural language processing. The transformer architecture uses attention mechanisms to process entire sequences simultaneously. This approach overcame limitations of sequential processing in traditional RNNs.

2.6.4 Current State of the Field

Deep learning continues to advance rapidly (Fig. 2.10).
Current research focuses on several key directions [38–40]:

- First, models have grown increasingly large and powerful. GPT (Generative Pre-trained Transformer) models can generate coherent text across various topics and styles. These models contain billions of parameters and demonstrate surprising capabilities in language understanding and generation.
- Second, researchers work to make deep learning more efficient. Techniques like knowledge distillation, pruning, and quantization reduce the computational requirements without sacrificing performance.
- Third, explainability has become a major focus. As AI systems make critical decisions, the need to understand their reasoning grows more important. Techniques for interpreting deep neural networks continue to develop.
- Fourth, AI ethics and bias mitigation receive increasing attention. Researchers recognize that models trained on historical data may perpetuate or amplify exist-

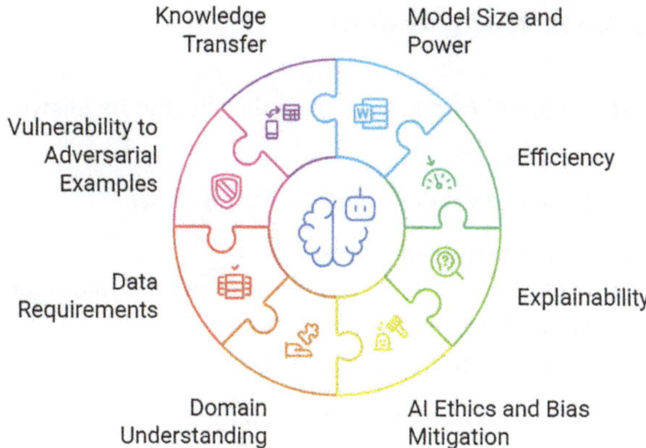

Fig. 2.10 Current Trends and Challenges in Deep Learning

ing biases. Methods to detect and mitigate bias have become an active area of research.

Several challenges remain for deep learning [8, 9]:

- Most systems still lack true understanding of their domains;
- Deep learning requires large amounts of labeled data;
- Systems remain vulnerable to adversarial examples;
- Transfer of knowledge between domains remains difficult.

Despite these challenges, deep learning has transformed expectations about artificial intelligence. Tasks once considered decades away have become reality. The practical impact of AI grows daily across industries and domains.

The deep learning era demonstrates the power of revisiting old ideas with new resources. Many core concepts existed long before their practical realization. This pattern suggests the value of maintaining diverse approaches in artificial intelligence research.

As deep learning continues to evolve, it connects with other AI approaches including symbolic reasoning, reinforcement learning, and cognitive modeling. These connections may address current limitations and lead to more general artificial intelligence systems.

2.7 Practice and Exercises

This section provides exercises to reinforce your understanding of AI history. The activities will help you connect key events, analyze critical developments, and appreciate the evolving nature of artificial intelligence.

2.7.1 Self-Assessment Questions

Test your understanding of key concepts from this chapter by answering the following questions:

1. Which of the following best describes the Turing Test?

 (a) A test measuring computational speed
 (b) A method for determining if a machine can exhibit intelligent behavior
 (c) A benchmark for neural network performance
 (d) A technique for programming early computers

2. When did the Dartmouth Conference take place?

 (a) 1943
 (b) 1950
 (c) 1956
 (d) 1970

3. Who is not considered one of the "founding fathers" of AI?

 (a) John McCarthy
 (b) Marvin Minsky
 (c) Alan Turing
 (d) Claude Shannon

4. Which period is known as the "first AI winter"?

 (a) 1956–1960
 (b) 1966–1973
 (c) 1974–1980
 (d) 1987–1993

5. What significant AI achievement occurred in 1997?

 (a) ELIZA passed the Turing Test
 (b) DeepBlue defeated world chess champion Garry Kasparov
 (c) Watson won the Jeopardy! championship
 (d) AlphaGo beat the world champion in Go

2.7.2 Answer Key

(b) A method for determining if a machine can exhibit intelligent behavior
(c) 1956
(c) Alan Turing
(c) 1974–1980
(b) DeepBlue defeated world chess champion Garry Kasparov

2.7.3   *Chronology Exercise*

Arrange the following key events in AI history in chronological order. For each event, write a brief explanation (1–2 sentences) of its significance:

(a) Development of ELIZA
(b) The Dartmouth Conference
(c) Publication of Turing's paper on computing intelligence
(d) First AI winter
(e) Creation of DeepBlue
(f) Rise of deep learning
(g) McCulloch-Pitts neural network model
(h) Expert systems boom
(i) AlphaGo defeats the world champion
(j) Emergence of machine learning approaches

This exercise will help you develop a clear timeline of AI development and understand how each event influenced subsequent advancements.

2.8   Conclusions and Future Directions

The history of artificial intelligence reveals distinct patterns across its developmental timeline. AI has progressed through cyclical phases of enthusiasm and disappointment. Each cycle has produced valuable insights and technical advancements.

Several key themes emerge from this historical analysis. The field has oscillated between symbolic and statistical approaches. Computational resources have consistently limited or enabled theoretical possibilities. Focused applications have generally outperformed attempts at general intelligence.

This historical perspective offers practical guidance for future development. Researchers should maintain realistic expectations based on previous cycles. Methodological diversity remains essential for addressing complex problems. Ethical considerations must precede widespread application. Interdisciplinary collaboration continues to drive innovation.

Deep learning currently demonstrates both remarkable capabilities and clear limitations. Issues of explainability, data efficiency, and robustness present ongoing challenges. Promising directions include hybrid approaches that combine neural methods with symbolic reasoning.

References

1. Wilson, R., Campbell-Kelly, M.: Computing: The 1940s and 1950s. Math Intelligencer. 42, 92–92 (2020). https://doi.org/10.1007/s00283-020-10009-x.
2. Peirce, C.S.: Collected Papers. Harvard University Press (1931).

3. Shannon, C.E.: Communication theory of secrecy systems. The Bell System Technical Journal. 28, 656–715 (1949). https://doi.org/10.1002/j.1538-7305.1949.tb00928.x.

4. Shannon, C.E.: A mathematical theory of communication. The Bell System Technical Journal. 27, 379–423 (1948). https://doi.org/10.1002/j.1538-7305.1948.tb01338.x.

5. TURING, A.M.: I.—COMPUTING MACHINERY AND INTELLIGENCE. Mind. LIX, 433–460 (1950). https://doi.org/10.1093/mind/LIX.236.433.

6. Oppy, G., Dowe, D.: The Turing Test. In: Zalta, E.N. (ed.) The Stanford Encyclopedia of Philosophy. Metaphysics Research Lab, Stanford University (2021).

7. McCulloch, W.S., Pitts, W.: A logical calculus of the ideas immanent in nervous activity. Bulletin of Mathematical Biophysics. 5, 115–133 (1943). https://doi.org/10.1007/BF02478259.

8. Russell, S., Norvig, P.: Artificial Intelligence: A Modern Approach. Pearson, Hoboken (2020).

9. Norvig, P.: Artificial intelligence: a modern approach. Global edition. Pearson, Boston (2021).

10. Frana, P.L., Klein, M.J.: Encyclopedia of Artificial Intelligence: The Past, Present, and Future of AI. Bloomsbury Publishing USA (2021).

11. Marquis, P., Papini, O., Prade, H.: A Guided Tour of Artificial Intelligence Research: Volume III: Interfaces and Applications of Artificial Intelligence. Springer Nature (2020).

12. Newell, A., Simon, H.: The logic theory machine–A complex information processing system. IRE Transactions on Information Theory. 2, 61–79 (1956). https://doi.org/10.1109/TIT.1956.1056797.

13. Ernst, G.W.: GPS and Decision Making: An Overview. In: Banerji, R.B. and Mesarovic, M.D. (eds.) Theoretical Approaches to Non-Numerical Problem Solving. pp. 59–107. Springer, Berlin, Heidelberg (1970). https://doi.org/10.1007/978-3-642-99976-5_3.

14. Nilsson, N.J.: The Quest for Artificial Intelligence. Cambridge University Press (2009).

15. Weizenbaum, J.: ELIZA—a computer program for the study of natural language communication between man and machine. Commun. ACM. 9, 36–45 (1966). https://doi.org/10.1145/365153.365168.

16. Winograd, T.: Procedures as a Representation for Data in a Computer Program for Understanding Natural Language. (1971).

17. Rosenblatt, F.: Perceptron Simulation Experiments. Proceedings of the IRE. 48, 301–309 (1960). https://doi.org/10.1109/JRPROC.1960.287598.

18. Rosenblatt, F.: The Perceptron, a Perceiving and Recognizing Automaton Project Para. Cornell Aeronautical Laboratory (1957).

19. Hemmer, M.C.: Expert Systems in Chemistry Research. CRC Press (2007).

20. Miller, F.P., Vandome, A.F., McBrewster, J.: Lighthill Report. Alphascript Publishing (2010).

21. Buchanan, B.G., Duda, R.O.: Principles of Rule-Based Expert Systems. In: Yovits, M.C. (ed.) Advances in Computers. pp. 163–216. Elsevier (1983). https://doi.org/10.1016/S0065-2458(08)60129-1.

22. Lindsay, R.K., Buchanan, B.G., Feigenbaum, E.A., Lederberg, J.: DENDRAL: A case study of the first expert system for scientific hypothesis formation. Artificial Intelligence. 61, 209–261 (1993). https://doi.org/10.1016/0004-3702(93)90068-M.

23. McDermott, J.: RI: an Expert in the Computer Systems Domain. Presented at the AAAI Conference on Artificial Intelligence August 18 (1980).

24. Aggarwal, C.: The Backpropagation Algorithm. In: Aggarwal, C.C. (ed.) Neural Networks and Deep Learning: A Textbook. pp. 29–71. Springer International Publishing, Cham (2023). https://doi.org/10.1007/978-3-031-29642-0_2.

25. Rojas, R.: The Backpropagation Algorithm. In: Rojas, R. (ed.) Neural Networks: A Systematic Introduction. pp. 149–182. Springer, Berlin, Heidelberg (1996). https://doi.org/10.1007/978-3-642-61068-4_7.

26. Vapnik, V.N.: The Support Vector method. In: Gerstner, W., Germond, A., Hasler, M., and Nicoud, J.-D. (eds.) Artificial Neural Networks—ICANN'97. pp. 261–271. Springer, Berlin, Heidelberg (1997). https://doi.org/10.1007/BFb0020166.

27. Wegner, S.A.: Support Vector Machines. In: Wegner, S.A. (ed.) Mathematical Introduction to Data Science. pp. 189–208. Springer, Berlin, Heidelberg (2024). https://doi.org/10.1007/978-3-662-69426-8_14.

28. Kecman, V.: Support Vector Machines—An Introduction. In: Wang, L. (ed.) Support Vector Machines: Theory and Applications. pp. 1–47. Springer, Berlin, Heidelberg (2005). https://doi.org/10.1007/10984697_1.

29. Bishop, C.M., Bishop, H.: Deep Learning: Foundations and Concepts. Springer, Cham (2023).

30. Goodfellow, I., Bengio, Y., Courville, A.: Deep Learning. The MIT Press, Cambridge, Massachusetts (2016).

31. Krizhevsky, A., Sutskever, I., Hinton, G.E.: ImageNet classification with deep convolutional neural networks. Commun. ACM. 60, 84–90 (2017). https://doi.org/10.1145/3065386.

32. IBM Watson, https://www.ibm.com/watson, last accessed 2025/02/24.

33. AlphaGo, https://deepmind.google/research/breakthroughs/alphago/, last accessed 2025/02/24.

34. Ünsalan, C., Höke, B., Atmaca, E.: Convolutional Neural Networks. In: Ünsalan, C., Höke, B., and Atmaca, E. (eds.) Embedded Machine Learning with Microcontrollers: Applications on Arduino Boards. pp. 285–321. Springer International Publishing, Cham (2025). https://doi.org/10.1007/978-3-031-69421-9_13.

35. Krauss, P.: Recurrent Neural Networks. In: Krauss, P. (ed.) Artificial Intelligence and Brain Research: Neural Networks, Deep Learning and the Future of Cognition. pp. 131–137. Springer, Berlin, Heidelberg (2024). https://doi.org/10.1007/978-3-662-68980-6_14.

36. Mitrofanov, E., Grishkin, V.: Generative Adversarial Networks Quantization. Phys. Part. Nuclei. 55, 563–565 (2024). https://doi.org/10.1134/S1063779624030596.

37. Madan, S., Lentzen, M., Brandt, J., Rueckert, D., Hofmann-Apitius, M., Fröhlich, H.: Transformer models in biomedicine. BMC Med Inform Decis Mak. 24, 214 (2024). https://doi.org/10.1186/s12911-024-02600-5.

38. Pressman, S.M., Borna, S., Gomez-Cabello, C.A., Haider, S.A., Haider, C., Forte, A.J.: AI and Ethics: A Systematic Review of the Ethical Considerations of Large Language Model Use in Surgery Research. Healthcare. 12, 825 (2024). https://doi.org/10.3390/healthcare12080825.

39. Sterling, T., Anderson, M., Brodowicz, M.: Chapter 19—Machine Learning. In: Sterling, T., Anderson, M., and Brodowicz, M. (eds.) High Performance Computing (Second Edition). pp. 383–393. Morgan Kaufmann, Boston/Waltham (MA) (2025). https://doi.org/10.1016/B978-0-12-823035-0.00019-5.

40. Liu, Z., Gu, Z., Liu, P.: Chapter 5—Machine learning basics. In: Liu, Z., Gu, Z., and Liu, P. (eds.) Transportation Big Data. pp. 129–175. Elsevier (2025). https://doi.org/10.1016/B978-0-443-33891-5.00005-3.

Chapter 3
The Turing Test and Fundamental AI Concepts

Abstract This chapter examines the Turing Test as a framework for evaluating artificial intelligence. It analyzes the test's origins in Turing's 1950 paper and its lasting impact on AI research. The discussion covers the test's methodology and technical requirements. Three key capabilities receive particular attention: natural language processing, knowledge representation, and automated reasoning. The chapter addresses major criticisms, including Searle's Chinese Room argument. It explores modern adaptations that extend beyond text-based evaluation. The analysis connects theoretical concepts to practical implementation challenges. Readers gain insight into intelligence evaluation methods and the defining characteristics of AI systems.

3.1   The Turing Test: Historical Context and Development

The Turing Test represents a pivotal milestone in the history of artificial intelligence, establishing one of the first formal methods for evaluating machine intelligence [1]. Understanding its origins and development provides essential context for appreciating its continued relevance in modern AI research and development.

3.1.1   Alan Turing's Background and Contributions to Computing

Alan Mathison Turing (1912–1954) was a British mathematician, logician, and computer scientist whose work laid the foundation for modern computing and artificial intelligence [2, 3]. Born in London, Turing showed exceptional mathematical abilities from an early age. He studied at King's College, Cambridge, where he developed groundbreaking ideas on computability theory.

O. Kuznetsov, *Intelligent Systems: From Theory to Applications*, Cognitive Technologies, https://doi.org/10.1007/978-3-032-00044-6_3

During World War II, Turing made crucial contributions to cryptanalysis at Bletchley Park. His work on breaking the German Enigma code significantly shortened the war and saved countless lives. This experience with computational problem-solving influenced his later thinking about machine intelligence.

Turing's 1936 paper "On Computable Numbers" introduced the concept of a universal computing machine (now known as the Turing machine) [4]. This theoretical device could, in principle, compute anything that is computable. The Turing machine concept established the theoretical foundation for all modern computers and introduced the notion that thinking might be considered a form of computation.

After the war, Turing worked on early computer designs at the National Physical Laboratory and later at the University of Manchester. He developed one of the first designs for a stored-program computer, which formed the blueprint for modern computer architecture. Turing's vision extended beyond mere calculation machines to computers that could exhibit intelligent behavior.

3.1.2 The 1950 Paper "Computing Machinery and Intelligence"

In 1950, Turing published his landmark paper "Computing Machinery and Intelligence" in the philosophical journal Mind [5]. The paper began with a bold statement: "I propose to consider the question, 'Can machines think?'" Rather than debating the philosophical definitions of "machine" and "thinking," Turing proposed replacing this question with a more practical test.

The paper described what he coined "the imitation game," later renamed to the Turing Test. Turing described an environment where a human judge is in natural language conversation with two participants, either machine and human. If the judge is unable to reliably distinguish between machine and human by their response, the machine is said to have passed.

Turing deliberately chose to use only text to isolate intelligence's thinking components and remove physical ability. In so doing, Turing left intelligence's most vital components free of physical instantiation to better evaluate computational intelligence.

The paper also anticipated major objections to the idea of machine intelligence. Turing methodically addressed nine potential counterarguments, including theological objections, the "heads in the sand" objection, mathematical limitations, arguments from consciousness, and Lady Lovelace's objection regarding original creativity. His responses demonstrated remarkable foresight about the future development and philosophical questions surrounding artificial intelligence.

3.1.3   *The Emergence of the Turing Test as a Benchmark for AI*

Although Turing originally predicted that machines would pass his test by the year 2000, the Turing Test gradually evolved from a futuristic prediction into a concrete benchmark for artificial intelligence research. The test gained prominence in the 1950s and 1960s as computer scientists began serious work on creating intelligent machines [2, 3].

Early AI researchers like John McCarthy, Marvin Minsky, Allen Newell, and Herbert Simon often referenced the Turing Test when discussing their long-term goals [6–9]. The test helped frame research objectives and provided a conceptual target for the nascent field of artificial intelligence. It offered a tangible, understandable goal that researchers, funding agencies, and the public could grasp.

The first significant attempt to implement Turing's ideas came with Joseph Weizenbaum's ELIZA program in the 1960s [10]. ELIZA simulated a Rogerian psychotherapist by reflecting user statements back as questions. Though limited in its capabilities, ELIZA demonstrated how even simple pattern-matching techniques could create the impression of understanding in specific contexts.

In 1990, Hugh Loebner established the Loebner Prize, an annual competition based on the Turing Test [2, 3]. This competition offered monetary rewards for computer programs that could fool human judges in text-based conversations. The Loebner Prize transformed the theoretical Turing Test into a practical competition, though critics argued that it encouraged shallow techniques aimed at brief deception rather than genuine advances in AI.

The emergence of large language models in the early twenty-first century has renewed interest in the Turing Test. Programs like GPT (Generative Pre-trained Transformer) and similar systems have demonstrated increasingly sophisticated language capabilities that approach human-level performance in many conversational contexts [11]. These developments have sparked renewed debate about the adequacy of the Turing Test as an intelligence measure and its relevance to modern AI research.

Throughout its history, the Turing Test has served multiple functions: a philosophical thought experiment, a concrete research goal, a practical benchmark, and a cultural touchstone for discussions about artificial intelligence. Its enduring relevance stems from its elegant simplicity and its focus on language as a fundamental aspect of intelligence. Despite numerous criticisms and proposed alternatives, the Turing Test remains a central reference point in evaluating progress toward human-like artificial intelligence.

3.2   The Turing Test Methodology

The Turing Test provides a practical framework for evaluating machine intelligence without becoming entangled in philosophical debates about the nature of consciousness or thought. This section examines the test's methodology, including its original formulation, structure, and the key technical components necessary for its implementation.

3.2.1 Definition and Purpose of the Test

The Turing Test, in its essence, evaluates a machine's ability to exhibit intelligent behavior indistinguishable from that of a human [1]. Unlike attempts to define intelligence abstractly, the test takes a pragmatic approach by focusing on observable behavior. This operational definition sidesteps complex philosophical questions about the nature of intelligence and consciousness.

The primary intent of the test is to provide an objective criterion for determination of whether or not to classify a machine as "intelligent" in the lack of clear definitions for intelligence. Turing's opinion was if a machine is capable of successful simulation of human linguistic behavior, then it is rational to assign to it a form of intelligence.

The test makes intelligence a function, and not a structure, of similarity. In short, no machine needs to think in the same structure humans do—it only needs to provide equivalent outputs for equivalent in-puts. The function-oriented approach has shown to have influence in AI development, for it focuses attention on capabilities, and not mechanisms.

Turing proposed the test not merely as a theoretical benchmark but as a practical goal for the field of artificial intelligence. He predicted that by the year 2000, computers would be able to play the imitation game so well that an average interrogator would have no more than a 70% chance of making the correct identification after 5 min of questioning.

3.2.2 Original Formulation and Protocols

In his 1950 paper [5], Turing described the test through a thought experiment he called "the imitation game." The original game involved three participants (Fig. 3.1): a man (A), a woman (B), and an interrogator (C) who could be of either gender. The interrogator would remain in a separate room from the other two participants and communicate with them only through written messages.

The initial goal of this game was for the interrogator to determine which participant was the man and which was the woman. The man's objective was to cause the interrogator to make an incorrect identification, while the woman's goal was to help the interrogator. Both participants could employ any strategies they wished, including deception.

Turing then proposed a modification to this game. He suggested replacing the man with a computer and changing the objective. In this modified version, the interrogator's task was to determine which participant was the human and which was the

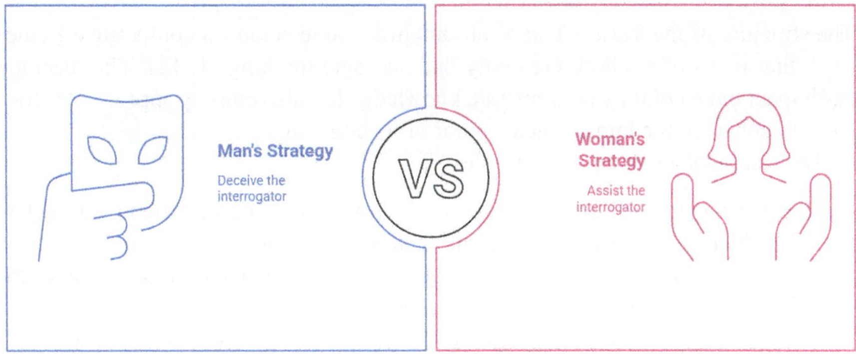

Fig. 3.1 The imitation game

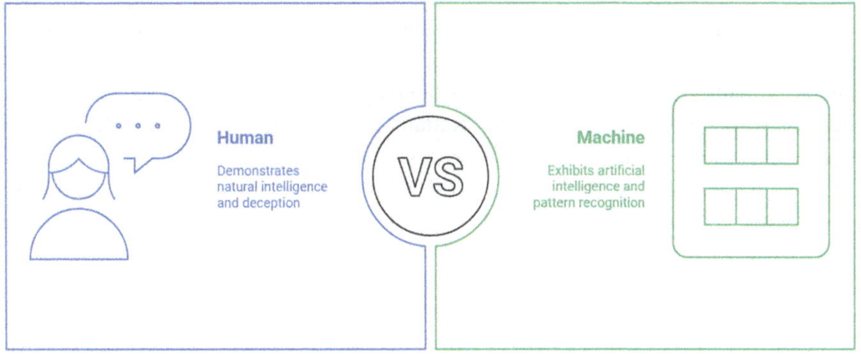

Fig. 3.2 The imitation game (modified version)

machine. If the interrogator could not distinguish between them reliably, the machine would be considered to have passed the test (Fig. 3.2).

The protocol for this modified game involved several key elements:

- Text-only communication to eliminate physical cues;
- Unrestricted topics of conversation to test broad intellectual capabilities;
- No limits on the interrogator's questions to prevent artificial constraints;
- A significant duration of interaction to test sustained performance;
- Statistical evaluation of success based on multiple trials.

Turing intentionally chose text-based communication to focus on cognitive aspects of intelligence rather than physical capabilities. This choice removed the need for physical embodiment, speech recognition, or computer vision, allowing the test to concentrate on the fundamental aspects of intelligence.

3.2.3 *The Imitation Game Structure*

The structure of the Turing Test as an imitation game creates a competitive framework that incentivizes both creativity and strategic thinking [1, 12]. The machine participant must not only demonstrate knowledge but also employ appropriate strategies to convince the human interrogator of its humanity.

The test involves three distinct roles:

- The Interrogator: A human evaluator who asks questions and attempts to determine which participant is human and which is a machine.
- The Human Subject: A person who answers the interrogator's questions truthfully, attempting to demonstrate their humanity.
- The Machine: A computer program designed to imitate human responses, with the goal of being indistinguishable from the human subject.

The interrogator can pose any question or initiate any conversation topic. This unrestricted format tests the machine's ability to handle unexpected queries, employ appropriate social conventions, exhibit common sense, and demonstrate general knowledge across domains (Fig. 3.3).

Success in the test is not binary but statistical. The machine's performance is evaluated based on the percentage of interrogators it can convince over multiple trials. A machine that consistently fools interrogators at a rate similar to that of a human participant (recognizing that humans might sometimes be misidentified as machines) would be considered to have passed the test.

The time duration of the test is an important factor. Turing initially suggested a five-minute interaction, but modern implementations often extend this period to

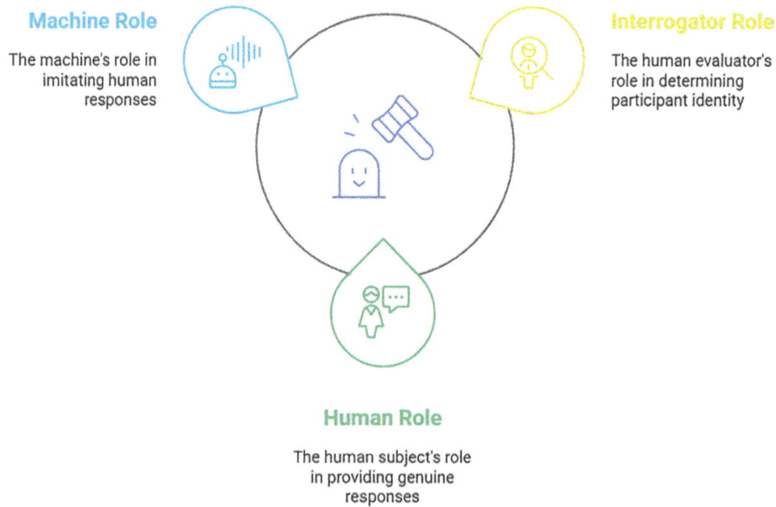

Fig. 3.3 Components of the Turing Test

provide a more thorough evaluation. Longer interactions test the machine's ability to maintain consistency and engage in extended reasoning.

3.2.4   Key Components: Natural Language Processing, Knowledge Representation, and Automated Reasoning

Successfully passing the Turing Test requires a machine to master three fundamental capabilities of artificial intelligence (Fig. 3.4):

1. Natural Language Processing (NLP) enables the machine to understand and generate human language. This involves:

   - Parsing and understanding the grammatical structure of questions;
   - Interpreting the semantic meaning behind words and phrases;
   - Recognizing contextual references and resolving ambiguities;
   - Generating grammatically correct and contextually appropriate responses;
   - Employing appropriate style, tone, and register for the conversation.

2. Knowledge Representation concerns how information is stored and organized within the system. A machine attempting the Turing Test must:

   - Maintain a comprehensive knowledge base about the world;
   - Organize facts into meaningful relationships and hierarchies;
   - Represent common-sense knowledge that humans take for granted;
   - Store information in a format that enables efficient retrieval and inference;
   - Update knowledge based on new information from the conversation.

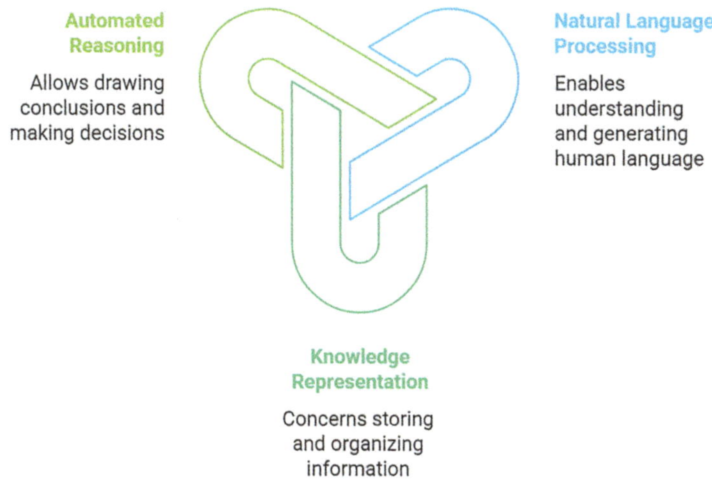

Fig. 3.4 AI's Path to Turing Test Success

3. Automated Reasoning allows the machine to draw conclusions and make decisions based on its knowledge. This includes:

- Making deductive inferences from known facts and rules
- Employing inductive reasoning to generalize from specific examples;
- Using abductive reasoning to generate the most plausible explanations;
- Managing uncertainty and probabilistic reasoning;
- Generating and evaluating hypotheses about the interrogator's intentions.

These three components must work together seamlessly to create the impression of human-like intelligence. The machine must process the interrogator's questions (NLP), access relevant knowledge (Knowledge Representation), reason about the appropriate response (Automated Reasoning), and then formulate a convincing reply (NLP again).

3.2.5   *Visual Representation of the Methodology*

The diagram above (Fig. 3.5) illustrates the structure of the Turing Test. The interrogator (C) communicates with both a human participant (B) and a machine (A) through a text-only interface. The interrogator cannot see either participant and must make a determination based solely on the content of their responses. The machine passes the test if the interrogator cannot reliably distinguish between the human and machine participants.

The diagram also highlights the three key AI components required for the machine participant: natural language processing (for understanding questions and generating responses), knowledge representation (for storing and organizing information), and automated reasoning (for drawing conclusions and making decisions).

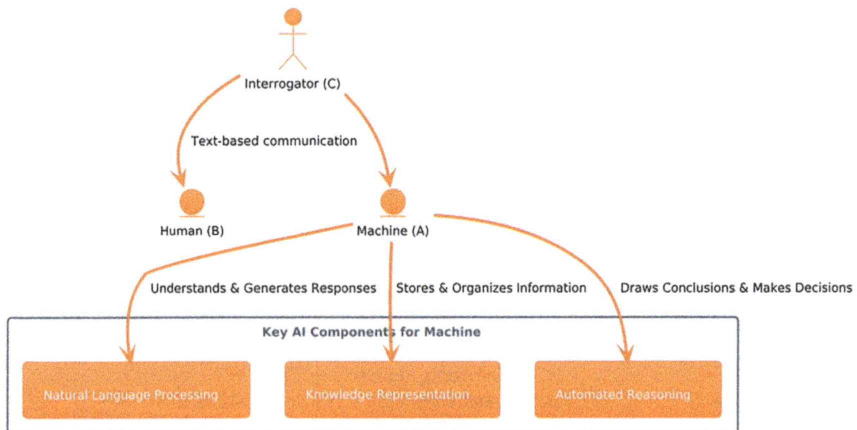

Fig. 3.5 Turing Test Method

These components work together to create a system capable of human-like conversation.

This methodology has proven remarkably durable as a conceptual framework for evaluating AI systems. Despite technological advances that have transformed computing since Turing's time, the fundamental challenge he identified—creating a machine that can engage in open-ended, natural language conversation indistinguishable from a human—remains at the frontier of artificial intelligence research.

3.3 Critiques and Limitations of the Turing Test

Despite its elegance and influence, the Turing Test has faced substantial criticisms from philosophers, computer scientists, and cognitive scientists. These critiques challenge both the test's methodology and its underlying assumptions about intelligence. Understanding these limitations provides a more nuanced perspective on the evaluation of machine intelligence.

3.3.1 Searle's Chinese Room Argument

The most famous critique of the Turing Test comes from philosopher John Searle's "Chinese Room" thought experiment, introduced in 1980 [13, 14]. Searle's argument directly challenges the assumption that passing the Turing Test demonstrates genuine understanding or intelligence.

In this thought experiment, Searle imagines himself locked in a room with a large rulebook and batches of Chinese writing. He does not understand Chinese. When Chinese symbols are passed into the room, he follows the rulebook's instructions to manipulate these symbols and pass different symbols back out. The rulebook is so comprehensive that his responses are indistinguishable from those of a native Chinese speaker.

From outside the room, it appears that whoever is inside understands Chinese. However, Searle argues that he still doesn't understand Chinese at all—he is merely following syntactic rules without any semantic understanding of what the symbols mean. By analogy, a computer program might pass the Turing Test by manipulating symbols according to programmed rules without genuinely understanding the conversation.

Searle distinguishes between:

- Syntax: The formal manipulation of symbols according to rules;
- Semantics: The meaning or understanding of what those symbols represent.

His argument suggests that computers operate purely at the syntactic level, while human intelligence involves semantic understanding. This distinction leads Searle to differentiate between (Fig. 3.6) [1, 12]:

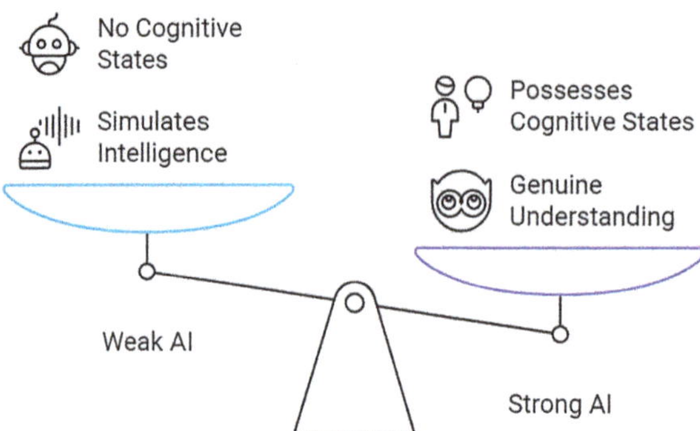

Fig. 3.6 Evaluating AI's Intellectual Claims

- Weak AI: The position that computers can simulate intelligence;
- Strong AI: The claim that computers with the right programs actually understand and have cognitive states.

According to Searle, passing the Turing Test might demonstrate weak AI but does not establish strong AI. Even a perfect simulation of understanding is not the same as actual understanding.

3.3.2 *Surface Behavior Versus Understanding*

Building on Searle's critique, many philosophers and AI researchers argue that the Turing Test evaluates only surface behavior rather than deeper cognitive processes. This "behavioral" approach to intelligence may be misleading for several reasons.

The test focuses exclusively on external outputs (responses to questions) without considering the internal processes that generate those outputs. Two systems might produce identical outputs through entirely different mechanisms—one through genuine understanding and reasoning, the other through sophisticated pattern matching or statistical correlations.

This limitation becomes particularly relevant with advances in large language models. These systems can generate remarkably human-like text by predicting likely word sequences based on statistical patterns in training data, without any underlying conceptual understanding of what they're discussing.

The distinction between performance and competence is crucial here. A system might perform well on the Turing Test without possessing the deeper competencies we associate with intelligence, such as (Fig. 3.7) [1, 12]:

- Causal reasoning about the physical world;
- Genuine belief formation;
- Intentionality (having thoughts about things);
- Consciousness or subjective experience;
- Self-reflection and metacognition.

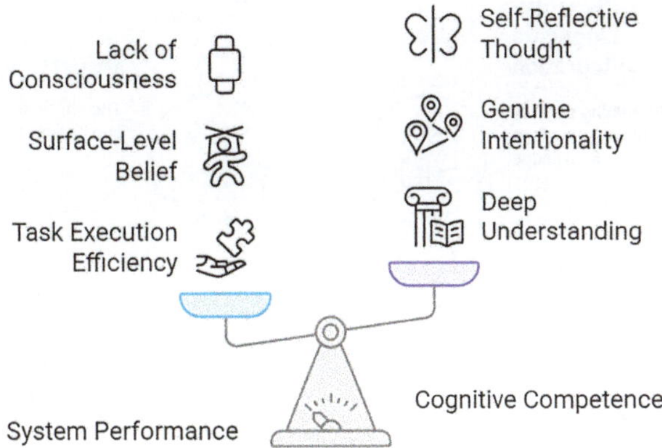

Fig. 3.7 Evaluating AI's True Cognitive Depth

This critique suggests that intelligence involves more than just producing appropriate outputs given certain inputs. Authentic intelligence requires internal representations, understanding, and genuine cognitive processes that the Turing Test cannot directly evaluate.

3.4   Modern Variants and Extensions

The original Turing Test has inspired numerous variations and extensions that address its limitations and expand its scope. These modern variants adapt the core methodology to evaluate different aspects of machine intelligence. Each variant introduces new dimensions to the assessment process, reflecting our evolving understanding of intelligence and the increasing capabilities of AI systems.

3.4.1   Total Turing Test

The Total Turing Test, proposed by philosopher Stevan Harnad in 1991 [15], significantly expands the evaluation criteria beyond text-based conversation. This variant recognizes that human intelligence encompasses physical interaction with the environment through perception and action.

In the Total Turing Test, the machine must demonstrate competence in sensory perception and motor capabilities alongside linguistic abilities. The machine would need to (Fig. 3.8):

- Process visual information and recognize objects in its environment;
- Understand and respond to auditory signals;
- Manipulate physical objects appropriately;

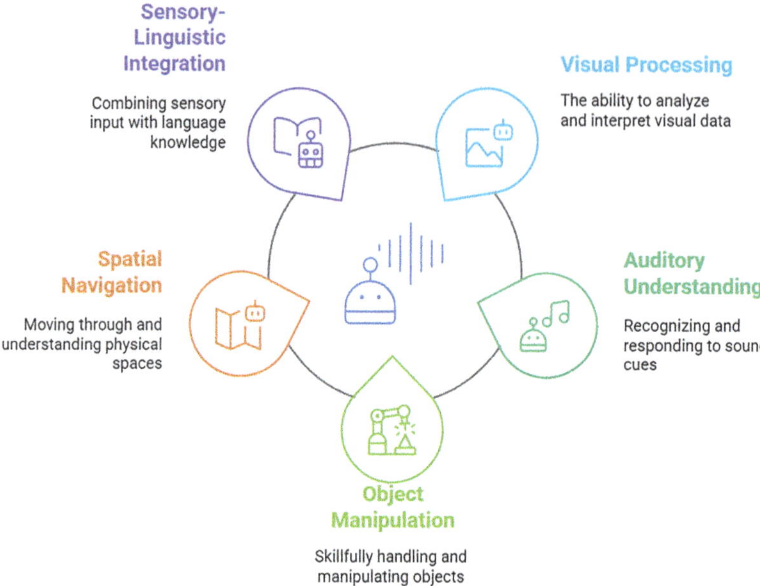

Fig. 3.8 Total Turing Test

- Navigate through physical space;
- Integrate sensory information with linguistic knowledge.

This extension addresses a fundamental limitation of the original test by acknowledging the embodied nature of human intelligence. Our cognitive abilities evolved in conjunction with our sensory and motor systems, and many aspects of intelligence relate directly to physical interaction with the world.

For a machine to pass the Total Turing Test, it would need to be embodied in a robotic form or have a virtual embodiment in a simulated environment. The interrogator would evaluate not only the machine's conversational abilities but also its performance on physical tasks and its responses to sensory stimuli.

The Total Turing Test aligns with research in embodied cognition, which emphasizes that thinking is deeply influenced by the physical body's interactions with its environment. This approach presents significant technical challenges, requiring integration of computer vision, robotics, and natural language processing. However, it offers a more comprehensive evaluation of intelligence that better reflects human cognitive capabilities.

3.4.2   Reverse Turing Test

The Reverse Turing Test inverts the roles of the original test [16]. Instead of a human trying to determine whether they are interacting with a machine or another human, the machine must determine whether it is interacting with a human or another machine.

This variant emerged partly in response to the proliferation of automated systems on the internet. Common implementations include:

- CAPTCHA (Completely Automated Public Turing test to tell Computers and Humans Apart).

- reCAPTCHA and similar systems that ask users to identify objects in images
- Behavioral analysis systems that distinguish human from automated browsing patterns.

The Reverse Turing Test highlights an important asymmetry: tasks that are trivial for humans but difficult for machines can serve as effective discriminators. Recognizing distorted text, identifying objects in cluttered images, or understanding cultural references often proves challenging for artificial systems while remaining straightforward for humans.

This variant holds practical significance for cybersecurity, helping protect online systems from automated attacks. It also serves as a reminder that intelligence involves recognizing intelligence in others—a form of meta-cognition that sophisticated AI systems would ultimately need to develop.

From a theoretical perspective, the Reverse Turing Test suggests that the ability to detect artificial behavior might be as important as the ability to generate natural behavior. A truly advanced AI system should be able to do both: act in ways indistinguishable from humans and recognize when others are not doing so.

3.4.3   Visual Turing Test

The Visual Turing Test extends evaluation beyond text to include visual processing capabilities [17]. This variant tests a machine's ability to understand and generate visual information in ways that match human performance.

In a typical implementation, the machine might be asked to (Fig. 3.9):

- Describe the contents of images in natural language;
- Answer questions about visual scenes;
- Generate images based on textual descriptions;
- Complete partial images in a contextually appropriate manner;
- Identify visual anomalies or inconsistencies.

One formal instantiation is the Visual Question Answering (VQA) task, where systems must answer natural language questions about images. These questions range from simple identification ("What color is the car?") to complex reasoning ("Why is the person smiling?").

The Visual Turing Test addresses the increasingly important role of visual intelligence in AI applications. Many practical AI systems, from autonomous vehicles to medical imaging analysis, require sophisticated visual processing capabilities alongside linguistic understanding.

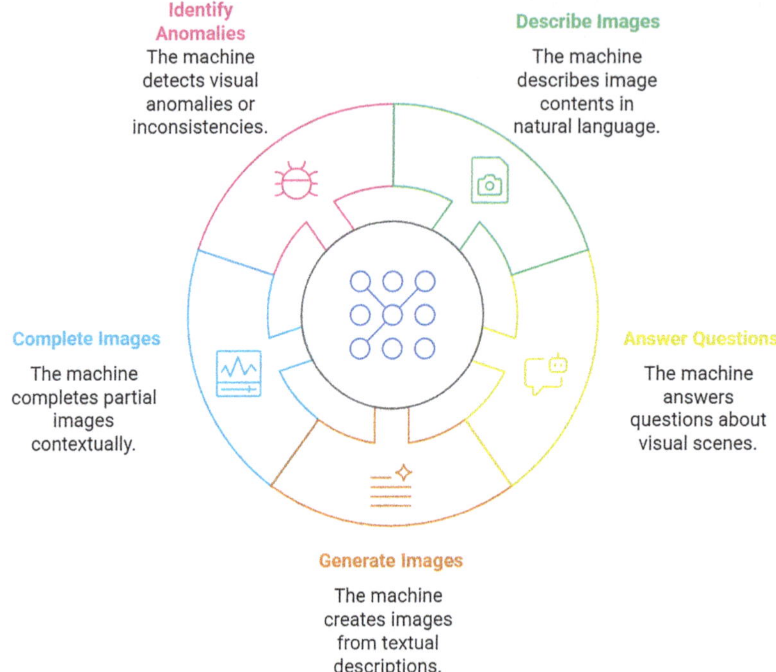

Fig. 3.9 Visual Turing Test

This variant acknowledges that human intelligence processes vast amounts of information visually. Our visual systems rapidly recognize objects, understand spatial relationships, interpret emotional expressions, and integrate this information with our linguistic knowledge—capabilities that advanced AI systems would need to replicate.

The Visual Turing Test represents a significant technical challenge, requiring integration of computer vision, natural language processing, and commonsense reasoning. Recent advances in deep learning, particularly in image recognition and generation, have made progress toward meeting this challenge, though fully human-like visual intelligence remains an ongoing research goal.

3.4.4   The Marcus Test for Real-World Reasoning

The Marcus Test, proposed by cognitive scientist Gary Marcus, focuses on evaluating machines' abilities to reason about the real world using common sense [1, 2]. This variant addresses a critical limitation of the original Turing Test, which can sometimes be passed through clever linguistic strategies without genuine understanding.

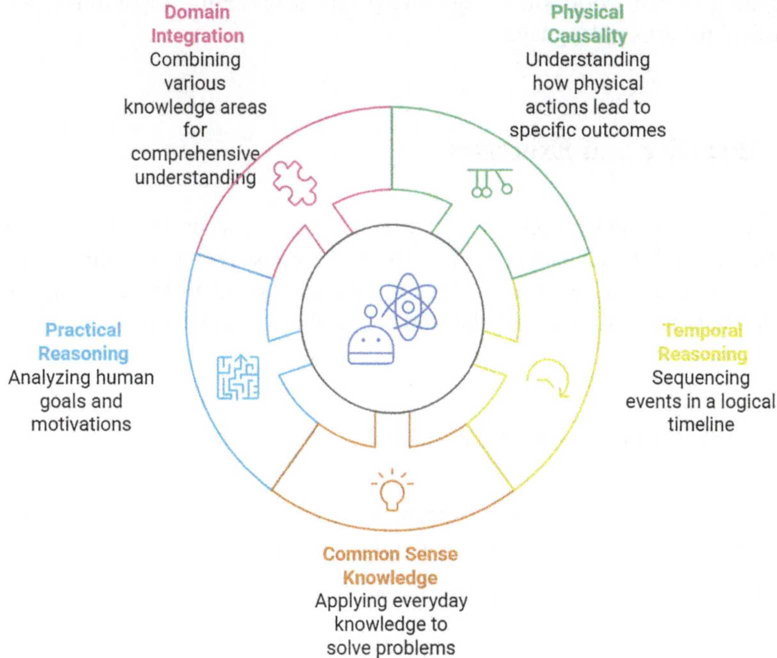

Fig. 3.10 The Marcus Test

The Marcus Test poses questions that require (Fig. 3.10):

- Understanding of physical causality;
- Temporal reasoning about sequences of events;
- Common sense knowledge about everyday objects and their properties;
- Practical reasoning about human goals, motivations, and behaviors;
- Integration of multiple knowledge domains.

For example, a question might be: "If I put a water bottle in the freezer, what will happen to the water?" Answering correctly requires understanding physical state changes, volume expansion, and the properties of containers.

Unlike the original Turing Test, the Marcus Test does not emphasize linguistic sophistication or conversational ability. Instead, it focuses on whether machines can demonstrate the kind of practical reasoning that humans use to navigate the physical and social world.

This approach identifies a crucial aspect of intelligence that current AI systems often lack. Large language models may generate fluent text that appears knowledge-able yet make basic reasoning errors about physical causality or practical conse-quences that no human would make.

The Marcus Test highlights the distinction between linguistic performance and genuine understanding. It suggests that true machine intelligence requires grounding in real-world knowledge and practical reasoning capabilities, not just statistical patterns in language.

3.5   Practice and Exercises

This section provides a range of activities to reinforce your understanding of the Turing Test and fundamental AI concepts. The exercises progress from basic comprehension questions to more complex critical thinking tasks, allowing you to develop both theoretical knowledge and practical analytical skills.

3.5.1   Self-Assessment Questions

Test your comprehension of key concepts covered in this chapter by answering the following questions. For each question, select the answer you believe is correct.

1. What primarily distinguishes the Turing Test from other intelligence evaluation methods?

   (a) It measures processing speed
   (b) It evaluates linguistic behavior without considering internal mechanisms
   (c) It requires specialized hardware
   (d) It focuses on mathematical abilities

2. Who proposed the "Chinese Room" argument as a critique of the Turing Test?

   (a) Alan Turing
   (b) John Searle
   (c) Marvin Minsky
   (d) Noam Chomsky

3. Which characteristic is NOT considered a fundamental feature of intelligent systems?

   (a) Learning capability
   (b) Adaptability
   (c) Processing speed
   (d) Reasoning ability

4. What does the term "strong AI" refer to?

   (a) AI systems with increased computational power
   (b) AI systems designed for specialized tasks
   (c) AI systems with genuine consciousness and understanding
   (d) AI systems that can perform multiple unrelated tasks

5. Which approach to AI focuses primarily on manipulating symbols and logical rules?

   (a) Connectionist approach
   (b) Symbolic approach
   (c) Machine learning
   (d) Deep learning

6. Which component of an intelligent system is responsible for making decisions based on available knowledge?

   (a) Knowledge base
   (b) User interface
   (c) Inference engine
   (d) Input/output module

7. Which critique of the Turing Test argues that external behavior does not necessarily indicate genuine understanding?

   (a) The anthropocentric bias critique
   (b) The superficiality critique
   (c) The Chinese Room argument
   (d) The linguistic limitation critique

8. In the context of machine learning, what is "supervised learning"?

   (a) Learning under the direct supervision of human experts
   (b) Learning from labeled examples with correct answers
   (c) Learning that requires constant monitoring
   (d) Learning from specially verified datasets

9. Which variant of the Turing Test includes evaluation of physical capabilities?

   (a) The Visual Turing Test
   (b) The Reverse Turing Test
   (c) The Total Turing Test
   (d) The Marcus Test

10. What was ELIZA, in the context of AI development?

   (a) The first AI to pass the Turing Test
   (b) A program simulating a psychotherapist using pattern matching
   (c) An expert system for medical diagnosis
   (d) A chess-playing AI

After completing these questions, compare your answers with the solution key provided at the end of this chapter. If you scored 8–10 correct, you have a solid understanding of the core concepts. If you scored 5–7 correct, review the relevant sections before proceeding. If you scored fewer than 5 correct, consider revisiting the entire chapter.

3.5.2　Solution Key for Self-Assessment Questions

(b) It evaluates linguistic behavior without considering internal mechanisms
(b) John Searle
(c) Processing speed
(c) AI systems with genuine consciousness and understanding
(b) Symbolic approach
(c) Inference engine
(c) The Chinese Room argument
(b) Learning from labeled examples with correct answers
(c) The Total Turing Test
(b) A program simulating a psychotherapist using pattern matching

3.5.3　Conceptual Understanding Exercise

Case Study: Medical AI Assistant

Consider an AI system called "MediAI" designed to assist doctors in diagnosing rare diseases. MediAI uses a vast database of medical records, research papers, and clinical data. The system interacts with physicians through natural language conversation, asking relevant questions, suggesting possible diagnoses, and recommending additional tests.

Task
- Identify at least three key characteristics of intelligent systems present in MediAI.
- Analyze how MediAI might succeed or fail in a Turing Test conducted in a medical context.
- Discuss two potential ethical implications of using such a system in medical practice.

Format
- Write a brief essay (250–300 words) addressing these three points. Structure your response with clear paragraphs corresponding to each point. Use specific examples to support your analysis.

Evaluation Criteria:

- Accuracy in identifying intelligent system characteristics;
- Depth of analysis regarding the Turing Test application;
- Relevance and thoughtfulness of ethical considerations;
- Clarity and coherence of expression.

3.6 Conclusions and Future Directions

The Turing Test represents a landmark contribution to artificial intelligence. It established a behavior-based approach to intelligence evaluation. This operational framework continues to influence research methods and goals.

Key limitations of the test reveal important aspects of intelligence beyond conversation. Searle's Chinese Room argument distinguishes behavior from understanding. Other critiques highlight the importance of embodied cognition and common-sense reasoning. These perspectives have expanded our conception of true intelligence.

Technical challenges in passing the test have driven advances in core AI technologies. Natural language processing, knowledge representation, and reasoning systems have all benefited from this pursuit. These technologies now power a wide range of applications beyond academic research.

Modern language models demonstrate impressive conversational abilities. They can engage in nuanced discussions across diverse topics. However, significant limitations persist. Current systems struggle with consistent reasoning and factual accuracy. They lack genuine comprehension of the concepts they discuss.

This gap between surface performance and deeper understanding creates a pivotal moment in AI development. Systems appear intelligent in constrained contexts yet fall short of human-level capabilities. This tension has renewed important discussions about the nature of intelligence. It raises questions about whether current approaches can achieve truly intelligent systems.

References

1. Russell, S., Norvig, P.: Artificial Intelligence: A Modern Approach. Pearson, Hoboken (2020).
2. Frana, P.L., Klein, M.J.: Encyclopedia of Artificial Intelligence: The Past, Present, and Future of AI. Bloomsbury Publishing USA (2021).
3. Nilsson, N.J.: The Quest for Artificial Intelligence. Cambridge University Press (2009).
4. Turing, A.M.: On Computable Numbers, with an Application to the Entscheidungsproblem. Proceedings of the London Mathematical Society. s2-42, 230–265 (1937). https://doi.org/10.1112/plms/s2-42.1.230.
5. TURING, A.M.: I.—COMPUTING MACHINERY AND INTELLIGENCE. Mind. LIX, 433–460 (1950). https://doi.org/10.1093/mind/LIX.236.433.
6. McCulloch, W.S., Pitts, W.: A logical calculus of the ideas immanent in nervous activity. Bulletin of Mathematical Biophysics. 5, 115–133 (1943). https://doi.org/10.1007/BF02478259.
7. Newell, A., Simon, H.: The logic theory machine–A complex information processing system. IRE Transactions on Information Theory. 2, 61–79 (1956). https://doi.org/10.1109/TIT.1956.1056797.
8. Rosenblatt, F.: The Perceptron, a Perceiving and Recognizing Automaton Project Para. Cornell Aeronautical Laboratory (1957).
9. Kelley, H.J.: Gradient Theory of Optimal Flight Paths. ARS Journal. 30, 947–954 (1960). https://doi.org/10.2514/8.5282.

10. Weizenbaum, J.: ELIZA—a computer program for the study of natural language communication between man and machine. Commun. ACM. 9, 36–45 (1966). https://doi.org/10.1145/365153.365168.
11. Restrepo Echavarría, R.: ChatGPT-4 in the Turing Test. Minds & Machines. 35, 8 (2025). https://doi.org/10.1007/s11023-025-09711-6.
12. Norvig, P.: Artificial intelligence: a modern approach. Global edition. Pearson, Boston (2021).
13. Cole, D.: The Chinese Room Argument. In: Zalta, E.N. and Nodelman, U. (eds.) The Stanford Encyclopedia of Philosophy. Metaphysics Research Lab, Stanford University (2024).
14. Searle, J.R.: Minds, brains, and programs. Behavioral and Brain Sciences. 3, 417–424 (1980). https://doi.org/10.1017/S0140525X00005756.
15. Harnad, S.: Other bodies, other minds: A machine incarnation of an old philosophical problem. Minds and Machines. 1, 43–54 (1991). https://doi.org/10.1007/BF00360578.
16. Albury, W.R.: Claude Bernard: Rationalite d'une methode. Pierre Gendron. Isis. 87, 372–373 (1996). https://doi.org/10.1086/357537.
17. Geman, D., Geman, S., Hallonquist, N., Younes, L.: Visual Turing test for computer vision systems. Proceedings of the National Academy of Sciences. 112, 3618–3623 (2015). https://doi.org/10.1073/pnas.1422953112.

Chapter 4
Modern Applications of Intelligent Systems

Abstract This chapter examines the diverse applications of intelligent systems across multiple sectors. It presents a structured overview of how AI technologies transform industries and services. The discussion begins with fundamental application categories and progresses to specific implementations. These include recommendation systems, virtual assistants, computer vision, autonomous vehicles, healthcare diagnostics, and financial analytics. The material balances technological foundations with societal implications. Through case studies and ethical analyses, readers gain practical understanding of real-world AI deployments.

4.1 Overview of Application Categories

Intelligent systems have permeated virtually every sector of modern society and economy [1, 2]. Their widespread adoption reflects both technological maturation and recognition of their transformative potential [3, 4]. This section provides a foundational understanding of the major application categories of intelligent systems, examining how they are classified and implemented across various domains [2].

4.1.1 Classification by Application Domain

Intelligent systems have various applications in disciplines, and every discipline is subject to unique needs and implementation plans. The discipline-specific applications of intelligent systems provide insights into where AI technologies fit in different contexts (Fig. 4.1).

In consumer services, automated and personal improvement of user experiences is achieved through smart systems. Recommendation systems suggest product, content, and services on the basis of behavior and preference patterns of the users. Online retail websites like Amazon use such systems to analyze web-browsing

O. Kuznetsov, *Intelligent Systems: From Theory to Applications*, Cognitive
Technologies, https://doi.org/10.1007/978-3-032-00044-6_4

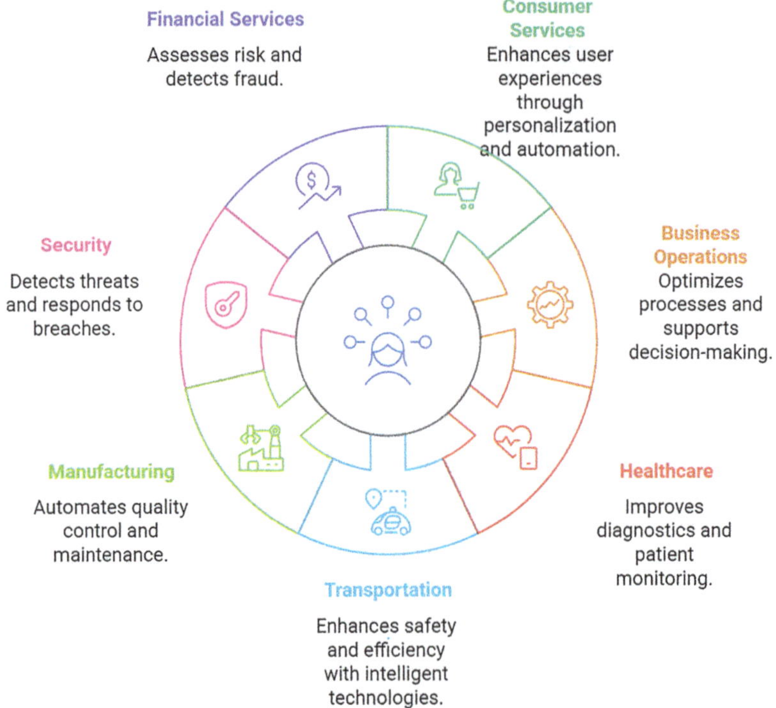

Fig. 4.1 AI Applications Across Domains

behavior and purchasing behavior, thus leading to enhanced improvement in satisfaction and conversion. Online content and music services like Netflix and Spotify have developed advanced recommendation systems to analyze watch and hear behavior to provide suggestions for content.

Business operations are augmented by smart systems by optimizing business decisions and business processes. The systems optimize resource planning, predict maintenance, and recognize areas of inefficiency by analyzing business operation data. For example, inventory systems use prediction analytics to maintain stock in optimal volumes, keeping stockout and excessive inventory cost to a minimum.

Healthcare applications demonstrate the life-saving potential of smart systems. Diagnostic aide tools analyze medical scans to recognize abnormalities that might otherwise go unnoticed by humans. Drug discovery systems accelerate candidate drug discovery by analyzing structures and predicting their interactions. Patient monitoring systems regularly check vital signs and alert medical staff to alarming changes, enabling timely interventions.

Transportation systems have started to rely on smart technologies for efficiency, safety, and self-driving. Smart vehicle technologies enhance vehicle safety through such advanced systems as automatic brake systems and lane departure warnings. Smart traffic systems receive current data and optimize signal timing and reduce congestion. Autonomous navigation systems blend sensor data and decision-making algorithms to allow for self-driving.

Manufacturing processes improve through smart systems in quality control automation, maintenance forecasting, and optimizing production. Products are inspected by computer vision systems in speeds and accuracy no labor by humans can match. Predictive maintenance software assesses equipment data to anticipate failures before they occur, reducing maintenance and downtime.

Security applications exploit smart systems for threat response and threat detection. Smart systems identify individuals in various contexts, such as building access and public safety. Anomaly detection systems identify deviant patterns, and such patterns may indicate intrusions in security. Network security systems monitor network activity constantly to identify and alert against potential intrusions in cyberspace.

Financial services employ smart systems for risk assessment, fraud detection, and automated trade. Credit score models analyze application data to predict repayment. Anti-fraud systems identify suspect patterns in real-time. Trading systems analyze market data in millisecond increments to execute optimal trades.

4.1.2   Classification by Technical Approach

Intelligent systems also fall into categories based on the underlying technical methods they use. They each have unique strengths that suit particular kinds of problem and application (Fig. 4.2).

Knowledge-based systems utilize explicit domain knowledge representation. They utilize rule engines, ontologies, and reasoning mechanisms based on logic to reason within a particular domain.

Machine learning models extract patterns and connections based on experience rather than rules programmed explicitly. Supervised models train with labeled instances and use them to make predictions with unseen instances. Unlabeled instances receive patterns and patterns within them with the use of unsupervised techniques. Reinforcement techniques construct the optimal behavior based on experience with the environment. They power a great deal of the technology that we utilize, including image recognition and natural language processing.

Computer vision systems receive visual world information and process and interpret it. They combine image processing and machine learning to recognize objects, understand scenes, and infer visual content.

Natural language processing technologies utilize human language either in speech or textual format. They leverage computational linguistics and machine learning methods to parse, generate, and translate human language.

Hybrid systems integrate multiple technical methods based on the strengths of each of them. Virtual assistants nowadays use natural language processing, knowledge representation, and machine learning to comprehend the queries of users and

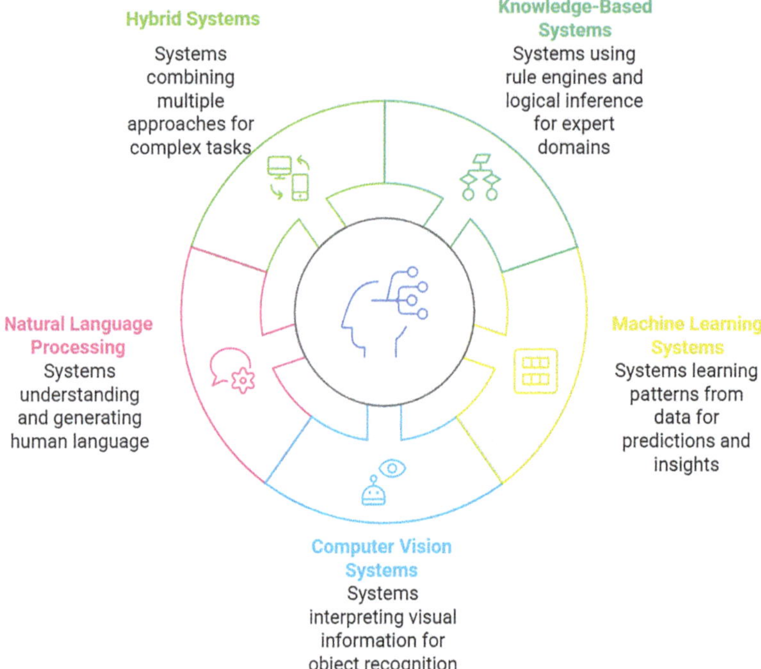

Fig. 4.2 Categorizing Intelligent Systems

respond suitably. Autonomous cars use computer vision, sensor fusion, and reinforcement learning to drive in complex environments in a safe manner.

4.1.3   Classification by Interaction Model

The way that users and environments are addressed by intelligent systems forms a useful taxonomy framework (Fig. 4.3).

Fully autonomous systems perform autonomously once set up, making and acting on decisions with no intervention required. They observe the environment they're in, take in information and perform the correct course of action based on the program and patterns they've acquired. Automated trading systems, autonomous automobiles in controlled environments, and specialized robots fall into this category. The greatest problem with such systems is making them highly reliable and safe in any situation they might encounter.

Collaborative systems aid humans and augment human abilities instead of supplanting them. The systems allocate tasks between human and machine based on each of their strengths. Robotic surgery aid improves the precision of the surgeon and reserves critical decision-making for human judgment. Robots that work alongside humans (cobots) perform the physically demanding and repetitive tasks and leave the

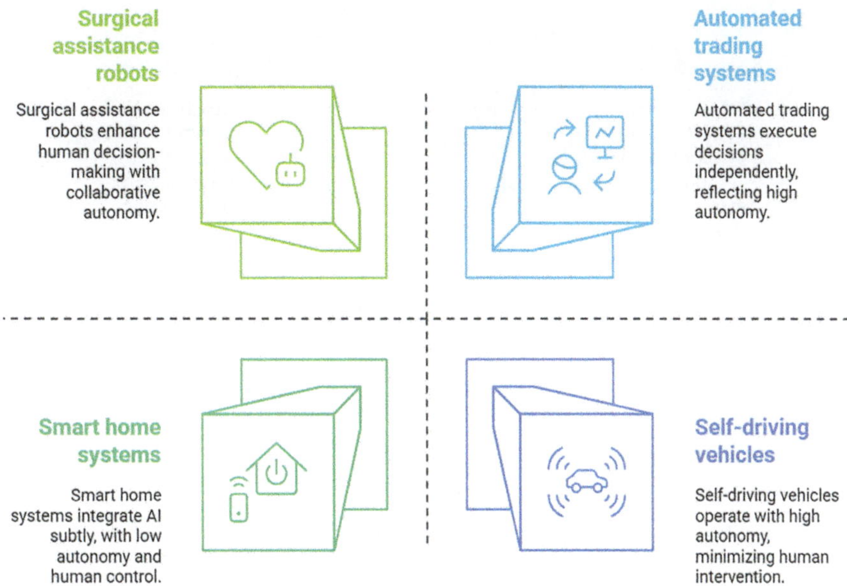

Fig. 4.3 Interaction Models of Intelligent Systems

complex tasks of production in the hands of the human workers. The systems need to be proficient at recognizing human intention and changing patterns of collaboration.

Decision support systems provide analysis and information that inform human decision-making but don't take any direct action. They transform complex data and present it in a comprehensible format. Medical diagnostic tools scan patient data and present suggestions of possible ailments but leave the final diagnoses to physicians. Financial planning guides scan the patterns of the market and the behavior of the portfolio and present investment suggestions. The power of such systems lies in analytical strength and interface design that match the thought processes of humans.

Embedded intelligence systems weave intelligence capabilities into the fabric of daily life and the environment. Smart home systems modulate environment states based on acquired preferences and identified patterns. Wearable physiological monitors scan physiological parameters continuously and offer insight into health. Smart infrastructure such as adaptive traffic lights adjust based on changing states. They work unseen behind the scene and gradually adjust based on patterns and preferences of the users.

4.1.4 Emerging Cross-Domain Applications

Many modern intelligent systems cross conventional domain boundaries and introduce novel applications that integrate capabilities drawn from multiple domains. Cross-domain applications typically stimulate great innovations through the application of proven techniques in a new environment (Fig. 4.4).

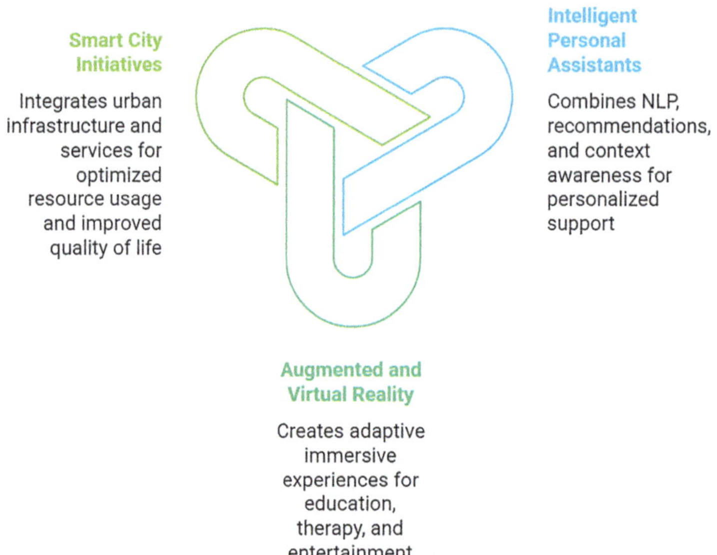

Fig. 4.4 Mapping Cross-Domain Intelligent Systems

Intelligent personal assistants take the strengths of several categories of capabilities and provide integrated support. They use natural language processing, recommendation systems, knowledge retrieval abilities and contextual intelligence. They provide highly individualized support based on users' preferences in categories including entertainment, purchase behavior, communications and productivity.

Augmented and virtual reality systems enhanced with intelligent capabilities create immersive experiences that respond adaptively to users. Educational applications can adjust difficulty based on learner performance. Therapeutic applications can tailor experiences to individual patient needs. Entertainment applications can generate dynamic content based on user preferences and reactions.

Smart city initiatives integrate intelligent systems across urban infrastructure, transportation, energy management, and public services. These interconnected systems share data and coordinate responses to optimize resource usage, improve service delivery, and enhance quality of life. Traffic management systems coordinate with public transportation and emergency services. Energy distribution networks communicate with building management systems to balance supply and demand.

The continued growth of such cross-domain applications is a testament to the plasticity of intelligent system technologies.

4.2   Consumer-Facing Applications

Consumer-facing applications of intelligent systems represent some of the most visible and widely adopted AI implementations (Fig. 4.5). These applications directly interact with end users, shaping their daily experiences with technology.

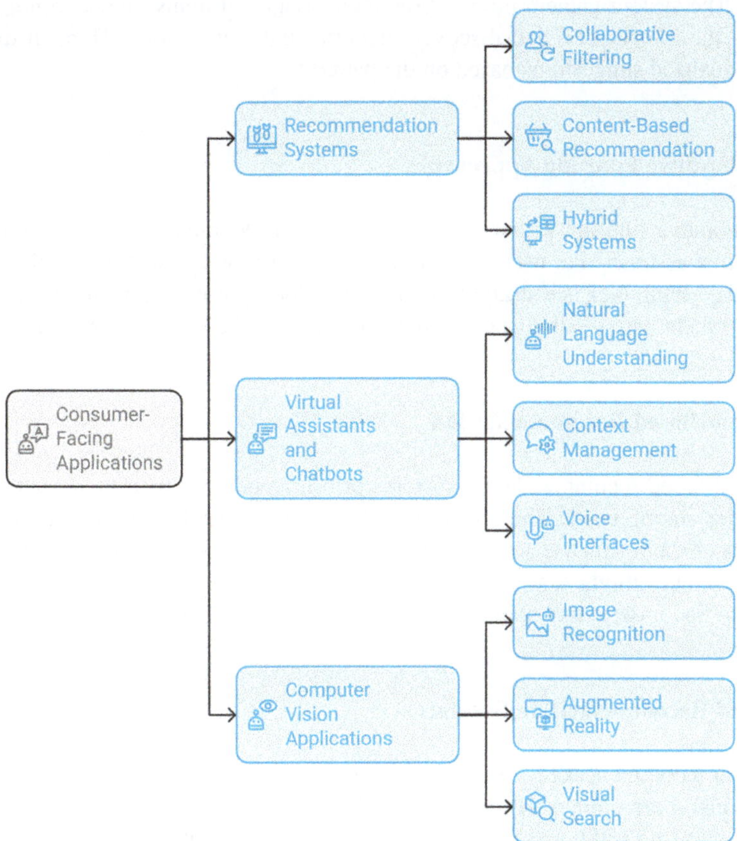

Fig. 4.5 Consumer-Facing Intelligent Systems Flowchart

4.2.1 *Recommendation Systems*

Recommendation systems pervade the internet and completely change the manner in which users discover goods, media, and services. They observe users' behavior and preferences and present them with the goods that they will be interested in. They influence engagement and satisfaction and commercial returns throughout the web.

Principles of Recommendation Engines

At their core, recommendation systems solve an information filtering problem. They filter relevant items out of large catalogs based either explicitly or implicitly on users' preferences. The overall recommendation process involves a few general steps. The system gathers information about users and items first. Second, it processes the information and discovers patterns and connections. Third, it delivers individualized suggestions based on the patterns.

Collaborative Filtering Approaches

Collaborative filtering is one of the widely used recommendation methods. The algorithm assumes that users who agreed in the past will also agree in the future. The underlying idea is a straightforward one: identify users with similar tastes and present them with the things that a specific user has liked based on similar tastes.

Content-Based Recommendation

Content-based recommendation systems prioritize item attributes over patterns of behavior among users. They build complex item profiles based on item attributes and match them with users' preference profiles. An example of a system that recommends movies might scan genres, stars, directors, and plot characteristics and use them to find movies that match an established interest of a user.

Hybrid Recommendation Systems

Modern recommendation engines often use a combination of methods to surpass individual shortcomings. Hybrid models blend collaborative filtering, content-based techniques, and additional methods to offer stronger and better predictions in a variety of contexts.

Applications in Streaming and E-Commerce

Recommendation systems have changed the landscape of e-commerce with the improvement of product discovery and a boost in sales. They drive purchase behavior with customized homepage presentation, "customers also bought" suggestions, and email campaigns.

4.2.2 *Virtual Assistants and Chatbots*

Virtual assistants and chatbots are yet another top category of customer-consumable intelligent systems. They enable natural language human-computer interaction and provide assistance, information, and services in the format of a dialogue.

Natural Language Understanding and Generation

Natural language understanding forms the foundation of virtual assistants and chatbots. The technology extracts the intention and the meaning of the speech and processes the users' speech. Modern-day NLU pipelines include a small set of processing steps that transform raw text into a structured representation.

Context Management in Multi-Turn Conversational Dialogue

Managing context is one of the most difficult tasks of virtual assistant design. Simple answer systems don't require this; conversational agents do because they need the ability to track multiple contexts during the dialogue.

Voice Interfaces and Voice Processing

Voice interfaces bring opportunity and additional sophistication into virtual assistant design. They free the hands of the users but require highly sophisticated speech processing pipelines. The leading systems integrate a variety of technologies that convert speech into text, derive meaning, and generate speech.

Applications in Customer Service and Personal Assistance

Virtual assistants and chatbots have been applied widely in customer and individual support contexts. They promise 24/7 availability and scalability and increasingly conversational capabilities that not only help companies and organizations but also users and individuals alike.

4.2.3 Computer Vision in Consumer Applications

Computer vision technologies have quickly moved beyond specialized business use and into the fabric of daily life among consumers. They take visual information and use it to drive new kinds of interaction, play, and productivity.

Image Recognition Systems

Image recognition constitutes the backbone of the majority of computer vision tasks applied in the consumer space. They recognize objects, activities, scenes, and faces within visual images. Most modern methods use convolutional neural networks that have been pre-trained with large image collections.

Augmented Reality Apps

Augmented reality supplements the real world environment with virtual information and creates interactive mixed-reality experiences. The computer vision provides the underlying capabilities that enable AR systems to understand the real world environment and place virtual contents accurately.

Visual Search Technologies

Visual search allows users to query with images instead of text. The systems scan visual media and recognize objects, extract features and retrieve associated information or similar images based on the visual content. The technology comes in handy in instances where textual explanations fall short or are inconvenient.

Entertainment and Social Media Programs

Computer vision has changed social media and entertainment with imaginative capabilities, interactive experiences, and deeper analysis of the media. They demonstrate how visual intelligence can augment social interaction and human creativity.

Privacy and Ethical Issues

Consumer-facing computer vision technologies present critical issues of privacy and ethics. They handle potentially sensitive visual data and therefore create benefits and dangers for users alike. Responsible development involves the resolution of a few major issues.

4.3   Industrial and Infrastructure Applications

While consumer applications of intelligent systems often receive the most public attention, some of the most transformative implementations occur in industrial and infrastructure contexts (Fig. 4.6).

These applications optimize production processes, transform transportation systems, and enhance public infrastructure. This section explores how intelligent systems drive efficiency, safety, and sustainability across these critical domains.

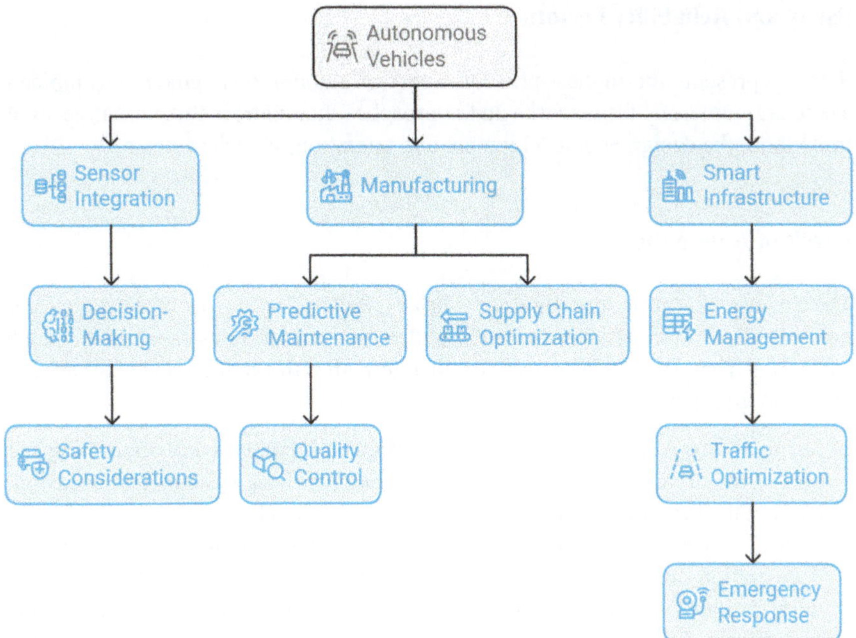

Fig. 4.6 Intelligent Systems in Industry and Infrastructure

4.3.1   Autonomous Vehicles and Transportation Systems

Autonomous vehicle technology forms part of the most ambitious application of intelligent systems. Autonomous systems leverage a combination of AI technologies to observe the world, reason and decide, and drive the automobile. Autonomous automobile development attempts to advance the safety, efficiency, and availability of transport.

Sensor Integration and Environmental Perception

Autonomous vehicles use a variety of sensor types to build detailed models of the environment. Different sensor types have different strengths and limitations. Efficient autonomous systems combine the varied sources of data to compensate for the limitations of individual sensors.

Decision-Making in Autonomous Navigation

Decision-making systems convert environment perception into vehicle behavior. This operation includes a set of integrated components that perform at a variety of timescales and abstraction layers.

Safety and Reliability Factors

Safety represents the highest-priority aspect of autonomous vehicle development. There are numerous frameworks and approaches that address this urgent necessity throughout the design and development and operation of such systems.

Levels of Automation

The Society of Automotive Engineers has established six levels of driver automation that offer a standard nomenclature with which autonomous capabilities may be addressed. They span the spectrum of no automation through complete automation in any environment.

- Level 0 (Zero Automation) has automobiles with human drivers performing each part of the drive. The automobile may include warning systems and occasional aid, but the driver has sole responsibility for the operation of the automobile.
- Level 1 (Driver Assistance) involves systems that aid in either acceleration/ deceleration or steering. Adaptive cruise control is a good example of this level with set speed and distance maintenance and the driver retaining responsibility for steering. Likewise, the driver controls position within a lane with the use of lane-keeping assistance and the driver remaining fully in charge of the operation of the entire vehicle at all times.
- Level 2 (Partial Automation) combines steering and acceleration/deceleration assistance. The vehicle can handle both lateral and longitudinal control in specific scenarios like highway driving. Many current production vehicles offer this capability, often marketed as "autopilot" or "co-pilot" systems. However, the human driver must maintain situational awareness, monitor the environment, and take control immediately when needed.
- Level 3 (Conditional Automation) enables the vehicle to handle all driving aspects under certain conditions. The human driver can disengage from actively monitoring the environment when the automated system is engaged. However, the system will request human intervention when encountering situations beyond its capabilities. The driver must remain available to take control when requested, presenting significant human factors challenges regarding attention readiness.
- Level 4 (High Automation) allows the car to drive whole trips with no intervention within a specific operational area. The area can be a specific geographic area, a set of roads of a specific type, or weather conditions. The car does the whole drive if it stays within the area and does not respond if a human driver does not answer intervention commands. The car does not drive autonomously outside the designated area.
- Level 5 (Full Automation) involves systems that perform any and all of the driver tasks in any situation that a human driver might be able to perform. They will drive any road and in any environment with no geographical or weather limita-

tions. They need no human driver capacity and potentially no controls at all. This level is still conceptual with major technical and legislative hurdles that need to be addressed prior to practical application.

Current commercial implementations typically remain at Levels 1 and 2 with limited instances of Level 3 and 4 systems in restricted environments. Level 4 operation has been initiated in geofenced environments with the application of such use cases such as autonomous delivery and robotaxis. The development through automation levels progresses in an evolutionary manner toward full autonomy with each phase possessing distinct technical and regulatory challenges.

4.3.2   Manufacturing and Predictive Maintenance

Intelligent systems have transformed manufacturing facilities with increased rates of automation, enhancement of quality and operation optimization. They leverage machine learning, computer vision, and prediction analytics in an effort to maximize productivity and lower downtime and costs.

Smart Factories and Industry 4.0

The concept of Industry 4.0 marks the fourth industrial revolution that encompasses the application of intelligent systems in the entire manufacturing processes. Smart factories live this vision with integrated systems that collect, analyze and take real-time decisions based on operational data.

Predictive Maintenance Systems

Predictive maintenance is among the greatest uses of intelligent systems within an industrial environment. The systems project failures in advance of the failures happening and allow maintenance at the best times. This reduces unplanned downtime and maximizes the life of the equipment and the use of maintenance resources.

Quality Control Automation

Intelligent systems revolutionized the procedures of the quality control with automatic fault detection and inspection. They provide reproducible and objective analysis within the mass production environment in which visual inspection becomes unreliable and inconvenient.

Supply Chain Optimization

Intelligent systems streamline the operation of the supply chain with superior forecasting, inventory management, logistics optimization, and risk management. They leverage the power of predictive analytics and optimization methods to optimize efficiency and resilience within complex supply networks.

Collaborative Robots and Automation

Collaborative robots, also known as cobots, mark an exciting development in automation technology. Traditional robots that move in individual safety cages no longer; instead, they move alongside human workers within a shared working environment. This collaborative strategy marries the problem-solving and flexibility of humans with the robotic strength and stamina.

4.3.3 Smart Infrastructure and Urban Systems

Intelligent systems increasingly optimize the management of infrastructure and the operation of cities. They use distributed sensing, analytics, and automatic controls to enhance efficiency, sustainability, and the quality of life within built environments.

Intelligent Energy Management

Energy infrastructure faces growing challenges from increasing demand, renewable integration, and sustainability requirements. Intelligent systems address these challenges through enhanced monitoring, prediction, and control capabilities across generation, transmission, distribution, and consumption.

Smart City Implementations

Smart city programs incorporate intelligent systems throughout the urban infrastructure, services, and operation. The implementations seek to optimize the quality of life, operational effectiveness, and the environment with the help of better monitoring, analysis, and management capabilities.

Traffic Optimization Systems

Traffic congestion costs major economies billions annually through wasted time, excess fuel consumption, and increased emissions. Intelligent transportation systems address these challenges through enhanced monitoring, analysis, and control of traffic flows.

Traffic monitoring systems provide real-time network insight. Inductive loops embedded within the roads monitor the occurrence and the speeds of the vehicles. The computer vision within video detection systems count the vehicles and record the speeds. Bluetooth and WiFi sensors track travel times based on anonymously identified device signatures at multiple locations. GPS-enabled floating car data provide network coverage beyond the positions of the fixed sensors. The multiple sources of the data provide network-wide visibility.

Emergency Response Coordination

Emergency management systems consolidate the use of resources and information throughout events that cross the spectrum of routine emergencies and catastrophic events. Smart systems enhance such capabilities with superior situation awareness, resource management, and decision-making support.

4.4 Specialized Domain Applications

Beyond consumer interfaces and industrial systems, intelligent systems have transformed specialized professional domains (Fig. 4.7). These applications leverage AI capabilities to enhance expert work in fields with distinct technical requirements and domain-specific knowledge. This section explores how intelligent systems address unique challenges in healthcare, financial services, and scientific research.

4.4.1 Healthcare and Medical Applications

Healthcare represents a highly promising and rapidly advancing field of application of intelligent systems. They augment the clinician's capabilities, diagnostic skills, and operational efficiency during the provision of healthcare.

Diagnostic Assistance Systems

Diagnostic assistance systems support clinicians in making diagnoses based on patient data. They contrast with patient-focused consumer health applications in that they integrate within the clinician workflow and enhance expert clinician judgment rather than replace it.

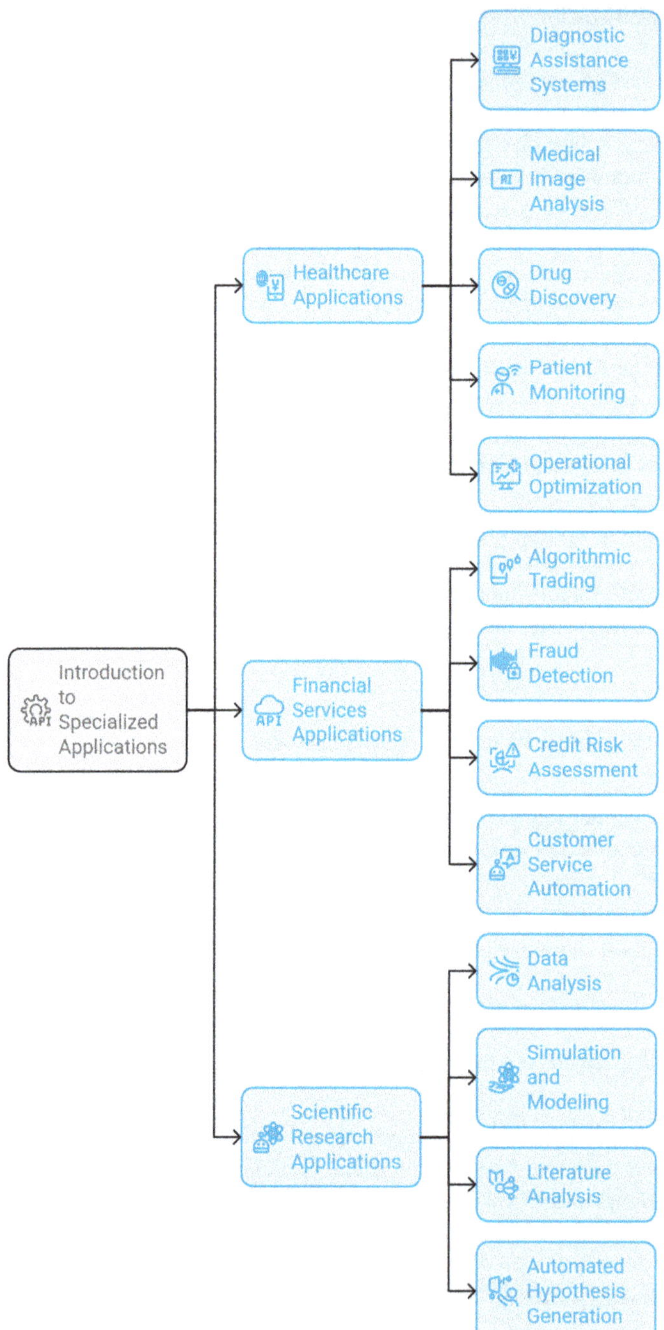

Fig. 4.7 Intelligent Systems in Specialized Domains

Medical Image Analysis

Medical imaging has experienced particularly rapid advancement through intelligent systems. Beyond basic diagnostic assistance, these applications enable quantitative analysis, longitudinal comparison, and procedure planning that enhance clinical capabilities.

Drug Discovery and Development

Pharmaceutical research and development faces significant challenges including high costs, lengthy timelines, and low success rates. Intelligent systems address these challenges by accelerating research, optimizing experimental design, and identifying promising drug candidates with greater efficiency.

Patient Monitoring Systems

Patient monitoring systems collect, analyze, and respond to physiological data in acute and outpatient settings. Smart capabilities enhance such systems with the detection of meaningful patterns, the prediction of critical events, and the enabling of clinician decision-making.

Administrative and Operational Optimization

Beyond direct clinical applications, intelligent systems enhance healthcare operations through administrative automation, resource optimization, and workflow improvements. These applications improve efficiency, reduce costs, and enhance access to care.

4.4.2   Financial Services Applications

Financial services offer an area highly amenable to the application of intelligent systems. The quantitative character of the associated data, the transparent metrics of performance, and the large economic rewards have promoted prompt and pervasive use throughout the sector.

Algorithmic Trading Platforms

Algorithmic trading uses computational methods of trading at rates and frequencies that surpass the capabilities of humans. Smart technology has elevated this field beyond simplistic automation into highly sophisticated strategies that incorporate detailed analysis of the market and flexible execution.

Fraud Detection and Security

Financial fraud constitutes a large and growing threat with global losses of hundreds of billions of dollars annually. Smart systems provide vital capabilities in the detection and deterrence of fraudulent activities throughout the payment system, account access, and monetary transactions.

Credit Risk Analysis

Credit risk analysis factors in the likelihood of the loans being paid in accordance with agreed-on obligations. Smart systems enhance this with superior application of the data, higher analysis methods, and real-time monitoring capabilities.

Customer Service Automation

Financial services involve numerous customer interactions ranging from simple transactions to complex advisory services. Intelligent systems automate routine interactions while enhancing human service delivery for more complex situations.

4.4.3   Scientific Research and Discovery

Scientific research increasingly employs intelligent systems to accelerate discovery, analyze complex data, and identify patterns beyond human analytical capabilities. These applications transform how scientists explore phenomena across disciplines from physics to biology and materials science.

Data Analysis of Large Scale Experiments

Modern scientific experiments produce record-breaking amounts of data beyond the limits of humans' abilities to review them. Particle physics experiments like the Large Hadron Collider produce petabytes of collision data. Sky surveys of the electromagnetic spectrum record tens of million stars and galaxies. Terabytes of a single run of DNA sequences are yielded by genomics tools. Smart systems help scientists extract useful insight from the large sets of data.

Simulation and Modeling

Simulation and modeling enable scientists to study events that are difficult or impossible to observe. Intelligent systems enhance such capabilities with superior optimization of parameters, identification of models, and analysis of uncertainty.

Literature Analysis and Integration of Knowledge

Scientific literature grows exponentially, with millions of new papers published annually across disciplines. Intelligent systems help researchers navigate this landscape through automated analysis, relationship discovery, and knowledge integration.

Automated Hypothesis Generation

Beyond literature review, computer systems play an active part in scientific discovery with the automatic formulation and testing of hypotheses. They also suggest new avenues of study based on patterns and connections that might not be detected by humans.

Case Studies in Physics, Biology and Materials Science

Concrete examples demonstrate how scientific discovery in a variety of fields is changed by intelligent systems. The case studies reveal the promise and the methods of AI-augmented scientific enquiry.

4.5 Practice and Exercises

This section provides hands-on exercises to reinforce your understanding of modern applications of intelligent systems. Through a combination of self-assessment questions, critical thinking exercises, group discussion topics, and practical case studies, you will develop both conceptual understanding and analytical skills related to AI applications across various domains.

4.5.1 Self-Assessment Questions

These questions test your understanding of key concepts related to modern applications of intelligent systems. For each question, select the answer you believe is correct. This will help solidify your knowledge and identify areas that may require additional review.

1. Which technique forms the foundation of modern recommendation systems in e-commerce platforms?

   (a) Computer vision
   (b) Natural language processing

(c) Collaborative filtering
(d) Reinforcement learning

2. In the context of virtual assistants, what does "intent recognition" refer to?

    (a) The ability to recognize human faces
    (b) Determining the purpose or goal of a user's query
    (c) Remembering previous user interactions
    (d) The capability to speak in different languages

3. Which technology is most critical for autonomous vehicle navigation in complex urban environments?

    (a) GPS positioning
    (b) Sensor fusion
    (c) Voice recognition
    (d) Facial recognition

4. What is the primary advantage of predictive maintenance systems in manufacturing?

    (a) They eliminate the need for human maintenance workers
    (b) They predict exactly when equipment will fail with 100% accuracy
    (c) They reduce unplanned downtime by identifying potential failures before they occur
    (d) They automatically order replacement parts without human approval

5. Which of the following best describes how modern AI-based diagnostic systems function in healthcare?

    (a) They completely replace human physicians in diagnosis
    (b) They primarily focus on administrative tasks like scheduling
    (c) They provide decision support to physicians based on pattern recognition
    (d) They only work with genetic data, not medical images

6. What is a "digital twin" in the context of industrial applications?

    (a) A backup copy of an artificial intelligence system
    (b) A virtual replica of a physical asset that mirrors its state and behavior
    (c) An identical copy of a dataset used for testing
    (d) A secondary robot that performs the same task as the primary one

7. Which of the following is NOT a common application of intelligent systems in financial services?

    (a) Fraud detection
    (b) Algorithmic trading
    (c) Physical security of bank vaults
    (d) Credit risk assessment

8. What is the main purpose of explainable AI in healthcare applications?

   (a) To make systems run faster
   (b) To provide understandable rationales for AI recommendations
   (c) To eliminate the need for medical training
   (d) To reduce the cost of healthcare

9. How do modern smart city applications typically use sensor data?

   (a) Solely for law enforcement purposes
   (b) Only for traffic management
   (c) For integrated decision-making across multiple city services
   (d) Exclusively for environmental monitoring

10. Which capability has most significantly enhanced the effectiveness of computer vision in consumer applications?

   (a) Faster internet connections
   (b) Deep learning techniques
   (c) Larger smartphone screens
   (d) Improved battery technology

4.5.2 Critical Thinking Exercise: Smart Transit System Analysis

Scenario

- The city of Metropolis has implemented an AI-powered public transportation system called "SmartTransit" to improve mobility and reduce congestion. The system integrates several intelligent components:
- Predictive algorithms forecast passenger demand based on historical patterns, weather, and special events
- Adaptive routing adjusts bus and shuttle routes based on real-time demand
- Dynamic pricing changes fares during peak hours to distribute demand
- Computer vision monitors passenger volumes at stations and on vehicles
- A mobile app provides personalized travel recommendations to users

After 6 months of operation, the system has reduced average wait times by 30% and increased ridership by 15%. However, several challenges have emerged:

  - Elderly residents report difficulty using the mobile app interface
  - Lower-income neighborhoods experience higher fare increases during peak hours
  - Some bus routes serving less populous areas receive reduced service
  - Privacy advocates express concerns about the computer vision monitoring system
  - Occasional algorithmic errors cause significant service disruptions

Task

- Analyze the SmartTransit system from multiple perspectives, considering both its benefits and challenges. Address the following points in your analysis:
- Identify the key intelligent system technologies employed in SmartTransit and how they contribute to its objectives
- Assess the ethical implications of the system, particularly regarding equity, privacy, and accessibility
- Suggest three specific improvements that could address the identified challenges while maintaining operational benefits
- Consider how principles from another application domain discussed in this chapter might be adapted to improve SmartTransit

Format your response as a structured analysis of approximately 400–500 words. Support your recommendations with clear reasoning and concepts from the chapter.

4.6 Conclusions and Future Directions

Modern applications of intelligent systems represent one of the most transformative technological developments of our time. These applications span virtually every domain of human activity, from consumer services to scientific discovery. Their continued evolution promises significant changes to how we live, work, and solve problems.

Several key insights emerge from our examination of applications across diverse domains:

First, successful implementations typically combine multiple AI capabilities rather than relying on single techniques. This integration creates more robust and effective systems than single-technique approaches.

Second, domain knowledge remains essential despite advances in general AI techniques. The most successful implementations balance technical sophistication with deep understanding of domain-specific requirements.

Third, human-AI collaboration often delivers better outcomes than either humans or AI systems operating independently. This collaborative approach leverages complementary strengths of both.

Fourth, implementation challenges extend beyond technical factors to include organizational, ethical, and social dimensions. Addressing these non-technical factors proves essential for real-world impact.

Finally, ethical considerations become increasingly important as intelligent systems influence more consequential decisions. Privacy protection, bias mitigation, transparency, and accountability represent critical concerns across all application domains.

References

1. Norvig, P.: Artificial intelligence: a modern approach. Global edition. Pearson, Boston (2021).
2. Frana, P.L., Klein, M.J.: Encyclopedia of Artificial Intelligence: The Past, Present, and Future of AI. Bloomsbury Publishing USA (2021).
3. Russell, S., Norvig, P.: Artificial Intelligence: A Modern Approach. Pearson, Hoboken (2020).
4. Nilsson, N.J.: The Quest for Artificial Intelligence. Cambridge University Press (2009).

Chapter 5
Problem Formulation and Search Spaces

Abstract This chapter explores problem formulation and search spaces in Artificial Intelligence. It presents essential theoretical foundations and practical approaches. The discussion covers formal problem definitions and representation techniques. Search space characteristics and navigation algorithms receive particular attention.

5.1 Foundations of Problem Formulation

Problem formulation represents the critical first step in applying artificial intelligence to solve complex challenges [1, 2]. This process transforms a real-world problem into a well-structured representation that enables algorithmic solutions [1]. When formulated properly, even highly complex problems become tractable through systematic application of search and optimization techniques [3, 4].

5.1.1 Formal Problem Definition

In the field of artificial intelligence, we can formally define a problem as a quintuple $P = (S, A, T, G, C)$, where each component plays a specific and essential role in the complete problem definition [3, 4]:

State Space (S): This represents the complete set of possible situations or configurations that can arise within the problem domain. Each element $s \in S$ represents a unique state of the problem. The state space may be discrete (finite or countably infinite) or continuous, depending on the problem's nature. For example, in a chess game, each possible arrangement of pieces on the board constitutes a unique state.

Action Set (A): This comprises all possible operations or moves that can be applied to transform one state into another. For a given state $s \in S$, the applicable actions may be represented as $A(s)$, the subset of actions that can be performed in state s. In the chess example, the actions would include all legal moves available to the player whose turn it is.

O. Kuznetsov, *Intelligent Systems: From Theory to Applications*, Cognitive Technologies, https://doi.org/10.1007/978-3-032-00044-6_5

Transition Function (T): The transition function $T : S \times A \rightarrow S$ defines how the system evolves when actions are applied to states. For a state ss s and an action $a \in A(s)$, the function $T(s, a)$ produces the resulting state s' after applying action aa a to state ss s. This function captures the dynamics of the problem environment.

Goal Test (G): The goal test $G : S \rightarrow \{true, false\}$ determines whether a given state satisfies the problem's objectives. For any state $s \in S$, $G(s) = true$ indicates that ss s is a goal state, while $G(s) = false$ indicates that further actions are needed to reach a solution. Some problems may have multiple goal states.

Cost Function (C): The cost function $C : S \times A \times S \rightarrow R^+$ assigns a numerical value to each state transition. For states ss s and s', and action aa a where $T(s, a) = s'$, the function $C(s, a, s')$ returns the cost associated with this transition. The cost function allows the evaluation and comparison of different solution paths.

This formal definition provides a precise mathematical framework that enables algorithmic approaches to problem-solving. By expressing a problem in terms of these five components, we transform an informal problem description into a structure that AI systems can process.

5.1.2   Key Components of Problem Formulation

Effective problem formulation requires detailed specification of each component. Let's examine how these components are developed in practice (Fig. 5.1):

Initial State Definition: The starting point of the problem has to be properly defined. The starting state $s_0 \in S$ is the state at which the solution search begins. The definition of this state has to be clear in order to ensure that the problem has a good foundation and that the solution search begins properly.

Action Specification: The set of possible actions must be comprehensively defined. Each action should have clear preconditions that determine when it can be applied and postconditions that describe its effects. Actions may be atomic or composite, depending on the problem's granularity. The action set directly influences the branching factor of the search space and, consequently, the computational complexity of the solution process.

Transition Model Building: The transition model reflects how the world progresses with the execution of an action. In deterministic environments, an action leads to a distinct predictable state. The transition function in the stochastic environment may incorporate probability parameters in an effort to encapsulate the uncertainties of the consequences. The transition model needs to accurately reflect the problem domain dynamics such that the state space will be traversed effectively by the search algorithm.

Goal Formulation Test: The definition of a satisfactory solution is stated in the goal formulation. This may be the realization of a specific state, the fulfillment of a set of constraint specifications, or the optimization of an objective function. The goal formulation needs to be computational-efficient since it will be evaluated

Fig. 5.1 Steps to Effective Problem Formulation

Cost Function Design

Design a function to quantify the desirability of solution paths.

Goal Test Formulation

Formulate what constitutes a successful solution.

Transition Model Development

Develop a model to describe world changes with actions.

Action Specification

Define possible actions with preconditions and postconditions.

Initial State Definition

Clearly specify the starting point of the problem.

numerous times during the search. The goal formulation of difficult problems may be a multiple criterion evaluation or admit a set of accept-able answers.

Cost Function Design: The cost function measures the desirability of solution paths. The cost function often reflects the resources spent (time, energy, dollars) or effort proxies such as distance traveled. The good cost function follows the real-world goals of the problem and directs the search toward the optimal or near-optimum solutions. The cost function has the power of dramatically affecting the efficiency of informed search methods.

Through deliberate formulation of such constituents, an actual problem can be recast into a properly formatted representation that will be solvable algorithmically. The transformation has a profound influence on the solution strategy that follows and how effectively and how efficiently it will be able to perform. Problem formulation faults will make easy problems computation-intensive and complex problems easy.

Iterative problem formulation refinement typically follows. Initial problem formulation will be revised with the emergence of a superior problem insight or recognition of computational limitations. The balance of abstraction needs to be correct—an excessive fine-grained detail will cause state space blowout and an excessive simplification will omit vital problem attributes.

5.2   State Representation Techniques

State representation plays a critical role in defining how effectively an AI system will be able to solve a problem [3, 4]. State representation is the particular technique applied to represent each of the problem do-main's possible states or configurations. The correct representation has the power to minimize computational resources and maximize solution efficiency. The wrong representation has the power to take a problem that might be easy and yet make it computationally infeasible.

Here we examine several ways of representing states in AI issues [1, 2]. We focus on representation techniques that try to balance computational efficiency and expressiveness. We seek to determine how the representation possibilities influence the problem-solving effort in general.

5.2.1   Importance of Efficient Representation

State representation influences a multitude of critical aspects of the problem-solving process in a direct manner (Fig. 5.2). Efficient representation reduces the computational effort of solution finding dramatically. Efficient representation also improves the algorithmic execution and extends the system's ability to identify the optimal solutions.

Representation has a dramatic influence on computational complexity. The state space doubles with each added variable or dimension of the representation. This has been labeled the "curse of dimensionality" and forces complex problems to be treated with an effective representation. The addition of each additional variable or attribute of the state representation has the power of duplicating the size of the search space.

Fig. 5.2 Components of Efficient State Representation

Memory also depends highly on representation efficiency. Memory becomes a limit in practical AI systems before computation power does. The states need to be stored during the course of the search and the memory usage per state has a direct effect on how many states might be searched. Efficient representation that prevents redundancy permits the dramatic enhancement of the depth of the search within memory bounds.

Algorithmic performance also depends on the structure that the representation imposes. Well-structured representation facilitates easy state transition and testing of the goals. They also enable the formulation of good heuristics that might guide the search algorithm. For example, a representation that preserves the natural problem structure might enable distance-based heuristics that accurately estimate the remaining distance to a goal.

The representation also affects the system's ability to utilize problem structure and recognize patterns. Good representation highlights useful regularities and patterns and brings them within the scope of the processes of reasoning and learning. This might be vital in complex environments if direct search becomes infeasible unless problem structure is utilized.

5.2.2 Atomic Vs. Factorial Representation

Typically, methods of state representation fall into two categories: atomic and factorial representation methods [1, 2]. The two methods each possess distinct advantages and limitations that fit them ideally to specific classes of problem (Fig. 5.3).

Atomic representation treats a state as an atomic unit. Here states are represented as individual and independent states with no state structure within them. The transitions between states are fully explained by the transition function. This representation typically includes a simple enumeration or state labeling. States may be labeled with a simple enumeration such as s_1, s_2, ..., s_n in a small state machine.

Atomic representation has the virtues of simplicity and directness. The states may be saved and compared with minimal use of a single identifier per state. The transition function may be a simple lookup table. This approach will be appropriate whenever the problem has a small set of states and the state structure does not play an important role in reasoning.

However, atomic representation becomes not feasible with a growing set of states. It fails to leverage state and problem structure similarities and does not scale effectively with large- and infinite-state space problems. The transition function needs to be programmed explicitly per state-action pair and becomes unmanageable with complex problem instances.

Factorial representation does fragment states into sets of variables or attributes. The state is a set of assignments of values to the variables. Considering that we may be having n variables and each of them has k values that it can take, the representation has the strength of representing at most k^n states effectively. One state within

Fig. 5.3 Comparison of
Atomic and Factorial
Representations

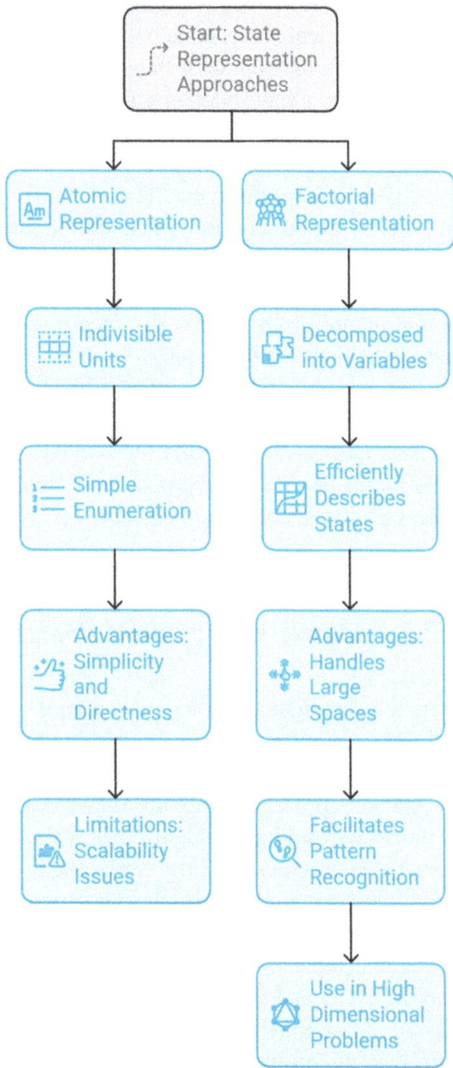

the representation can be notated as $s = (v_1, v_2, \ldots, v_n)$ and each of the variables v_i will be carrying the value of the i-th variable.

Factorial representation has the power of representing state spaces of unimaginably large sizes in a tight space. Factorial representation has the natural power of capturing the compositional character of complex problems and enabling the formulation of general transition functions and general heuristics. Factorial representation also enables the system to recognize patterns and similarities among states based on the similarities of the variables' values.

Factorial representation comes into play with highly dimensional issues. Factorial representation allows the transition function to be defined in terms of the effect of an action on individual variables rather than the state as a whole. This localized technique has the effect of making the problem definition and solution drastically simpler. For a problem of a navigating robot, the state might include position coordinates, direction of heading, battery power and sensor inputs—each affecting different aspects of state transition.

5.2.3   Representation in Classic Problems

Considering representation possibilities within typical AI problems provides real insights into state representation techniques [1, 2]. They illustrate how representation decisions influence solution strategies (Fig. 5.4).

The 8-puzzle problem is a good example of factorial representation. The 8-puzzle problem involves a 3×3 square with 8 numbered tiles and a blank space. The issue at hand is how the tiles will be moved such that they will be in a final state. The state of this problem has two representation forms: a 3×3 matrix that indicates the position of each tile and a vector of 9 that declares the contents of each of the cells (1–8 or blank).

For the 8-puzzle, a factorial representation could use variables v_1, v_2, ..., v_9 to represent the contents of each position, or variables representing the position of

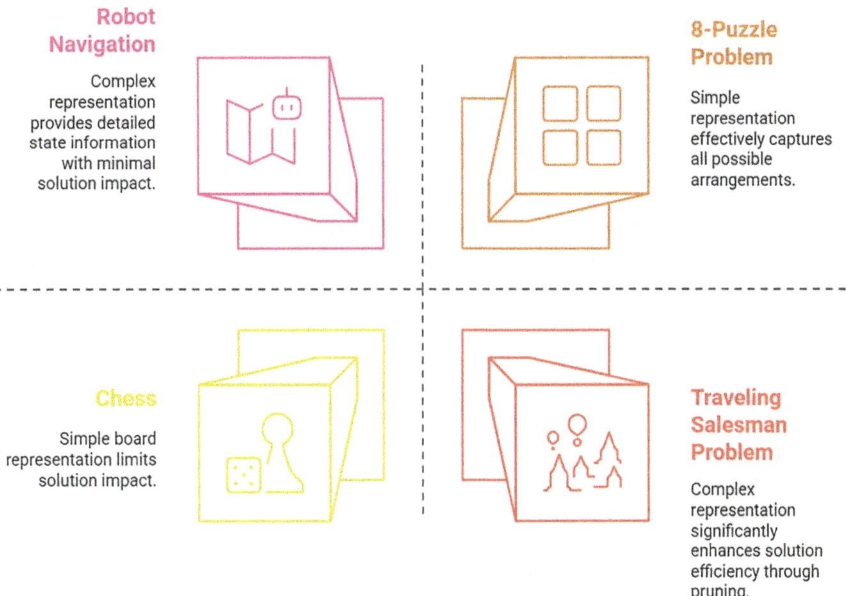

Fig. 5.4   Representation Strategies in Classic AI Problems

each tile. The two representation methods affect the state space size and the ease of specifying the operators. Under either representation, the system has the power of representing any of the 9 ! = 362, 880 configurations, but the representation permits transitions and the goal tests to be coded in a brief manner.

Travelling salesman problem (TSP) seeks the shortest route passing through each of the cities once and returning to the starting city. State representation of the TSP typically codes the partial tour built so far. Factorial representation might use variables that stand for the present city and the set of visited cities so far. Or states might be coded as permutations of the cities that stand for the visited cities' order.

In the problem of TSP, the representation has a direct effect on the branch factor and the application of branch methods. The representation that involves the cumulative sum of the cost so far allows branch-and-bound techniques that perform better. The fac-torial representation allows the system to recognize that the various partial tours have critical properties that may be the foundation of computation reuse.

Other standard issues illustrate further representation methods. In chess, boards are represented by states in chess games, normally with an 8×8 grid of cells each carrying state and color of the pieces. Position and direction of the state are normally coded in pathfinding issues such as robotic route finding. Sometimes extra state such as battery life and carried objects might be added.

These illustrations emphasize how representation options need to be compatible with the problem structure and the solution strategy that will be used. A representation that will allow easy state transitions, good development of heuristics, and use of problem structure will dramatically boost problem-solving proficiency.

5.3 Search Space Analysis

Search space analysis provides us with useful insight into how problem-solving strategies need to be built within artificial intelligence [3, 4]. The knowledge of a search space and its attributes enables us to select the appropriate algorithm, estimate computational requirements, and identify the source of the problem. This section investigates the underlying properties of the search space and how they affect problem-solving within AI.

5.3.1 Definition and Properties of Search Spaces

Search space forms the whole problem-solving environment an algorithm has to travel through in a bid to reach a solution (Fig. 5.5).

We mathematically define the search space as the set of states that we get if we start with the initial state s_0 and use the accessible actions. Algebraically, the search space Σ of a problem with an initial state s_0 might be mathematically defined as

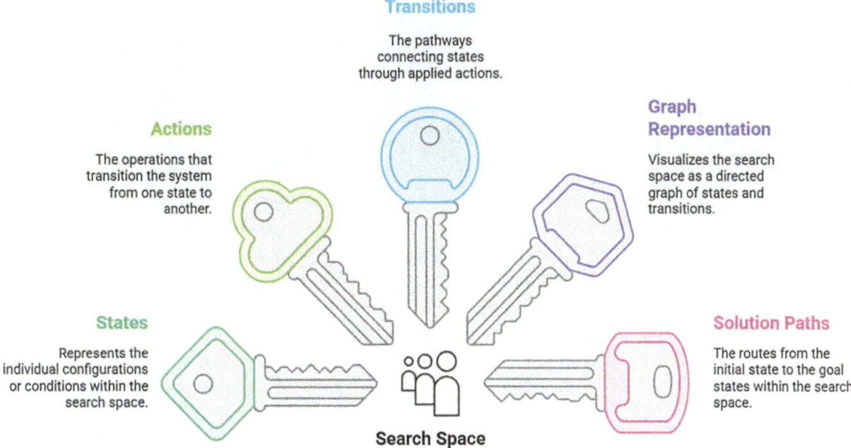

Fig. 5.5 Navigating Complex Landscapes

$$\Sigma = \left\{ s \in S \mid \exists a_1, \ldots, a_n \in A : s = T(\ldots T(T(s_0, a_1), a_2)\ldots, a_n) \right\}.$$

This definition encompasses all states that may be reached through the execution of some sequence of actions from the starting state. The space of the search space forms a subset of the state space because there may be states that theoretically exist but that cannot be reached starting with the provided initial state.

We can conceptualize the space of the search as a directed graph. Nodes will be states and arcs will be transitions that arise due to an action. The initial state will be the root of the graph and the goals will be the endpoints of the search. One of the tree paths leading toward a final state will be a solution of the problem that has the potential of working.

The ease of finding solutions will be affected directly by the shape of the search space. Well-structured search spaces that nudge the algorithm toward the solution will be easier to solve than complex and unstructured space. The scale and how the space has been set up will affect the probability that a specific strategy will be able to succeed within the bounds of practical computation.

5.3.2   Key Properties of Search Spaces

There are a few crucial characteristics that identify the space of search and influence the efficiency of solution techniques (Fig. 5.6). They reveal the natural problem difficulty and dictate the identification of appropriate algorithms.

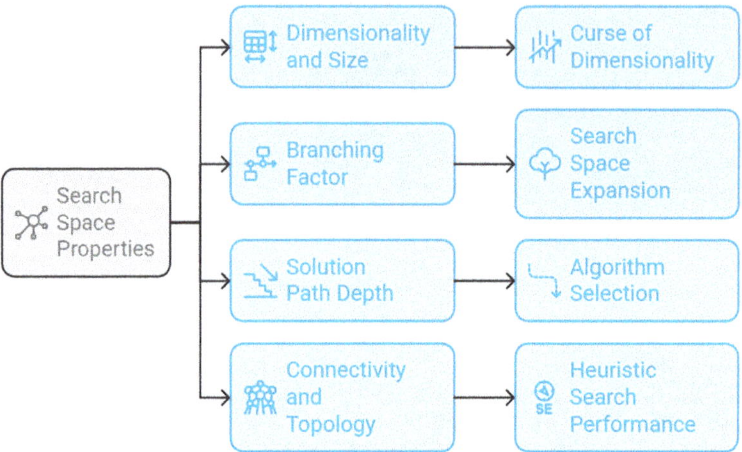

Fig. 5.6 Key Properties of Search Spaces

Size and Dimensionality of the State Space

The dimensionality of a state space is the number of attributes or variables it needs in order to define a state. The size of the state space, $|\Sigma|$, is the count of reachable states. Both of them have a great effect on computational demand.

Search space grows exponentially with dimensionality—the infamous "curse of dimensionality." The problem state space of a problem with n binary variables has a size of 2^n, for example. This exponential growth prevents exhaustive search of large-dimensional problem instances. Algorithms must use a heuristic or other strategy if the search space becomes so large that exhaustive search is not feasible.

Size of the search space also influences memory requirements. The frontier nodes or visited states must be stored by standard search al-gorithms. The system will often be impacted by memory limitations before the processing power with an expanding search space. This might compel algorithmic techniques that trade space efficiency against time efficiency.

Branching Factor

The branching factor b of the typical number of relevant actions per state captures the rate at which the state space grows with depth of the search. The higher the branching factor, the more difficult the breadth-first methods will be because the states that need to be searched will grow exponentially with depth.

We calculate the branching factor as the mean of the successors of all the states of the space of the search:

$$b = \frac{1}{|\Sigma|} \sum_{s \in \Sigma} |Successors(s)|.$$

We write *Successors(s)* for the set of states reachable with a single application of an action to state *s*.

In real life, the branching factor may be different in different areas of the search space. There may be states with numerous relevant actions and states with few relevant actions. The variation has a great influence on algorithmic performance and may imply specialized methods of search in different areas of the search space.

Solution Path Depth

The solution path depth of a problem, labeled as *d*, is the shortest number of steps required in order to reach a solution state based at the starting state. The minimal solution depth, also known as the d^*, dictates the shortest solution path depth. This attribute has a direct effect on the minimal solution finding time.

Solution depth also affects algorithmic selection. Deep solution paths put depth-first techniques at a disadvantage if they get trapped in the long explorations of the inappropriate paths. Breadth-first techniques might be at a disadvantage with highly branching issues with mod-erate solution depths.

In the majority of real-world problems, the depth might be unlimited or enormously large. Under the absence of any direction or bounds, the search programs may follow paths of unlimited depth with no solution found. Cycle detection and depth bounds play a crucial role in such instances to avoid infinite traversal.

Connectivity and Topology

The topological structure of a search space describes how states relate each other with respect to the transitions between them. This characteristic encompasses features such as the occurrence of dead ends (successors of states with no successors), plateaus (domains with states of similar evaluation values), and bottlenecks (narrow paths linking larger areas of the space).

Topology of the search space also has a great influence on the behavior of heuristic search techniques. Those with smooth gradients toward the goal states allow good heuristic directionality. Other landscapes with local maxima will steer the heuristic searches astray from the correct solution. Designing the correct heuristics and the appropriate strategy of search becomes easy with the knowledge of such topological attributes.

There are certain search spaces that contain disconnected areas that cannot be reached between each other. The solution in such a situation might be in a disconnected part of the initial state and the problem will be non-solvable. The detection of this situation at an initial phase prevents useless computational effort.

These factors contribute to characterizing the overall hardness of a particular space that needs to be searched. One problem with a large branching factor and short solution paths has a different set of issues compared with a problem with a small branching factor and highly deep solution paths. We can utilize these factors to drive algorithmic and resource-allocation choices based on the problem domain.

5.4   Search Graphs and Trees

Graphic models provide us with strong tools with which we can visualise and reason about search space [3, 4]. The visual forms enable us to understand the between-state connections and identify solution paths and design useful and effective algorithm designs [1, 2]. This section looks at two primary graphical models: search graphs and search trees and looks at their properties and algorithm design use.

5.4.1   Search Graphs: Structure and Properties

A search graph $G = (V, E)$ provides an explicit representation of a search space [1]. Under this representation each node $n \in V$ corresponds to a particular state of the search space and each directed edge $(n, n') \in E$ represents a state transition that results after an action has been applied. The search graph captures the whole structure of the search space within a representation that an algorithm may traverse.

It begins with a node representing the starting state. Edges move outward from this root node toward nodes representing states reachable with a single actions. This goes on with each node connected with successors with directed edges. The state reachable with multiple sequences of actions has a unique occurrence in the graph with multiple entering edges representing different paths leading toward that state.

Formally, for a problem with state space S, action set A, and transition function T, the corresponding search graph has:

- A vertex set $V \subseteq S$ containing all reachable states;
- An edge set $E = \{(s, s') \mid s, s' \in V, \exists a \in A : T(s, a) = s'\}$.

The topological structure of the search graph has a direct representation of the crucial attributes of the underlying problem. Cycles within the graph imply that the states may be revisited with alternate sequences of actions. Nodes that are dead-ends identify states with no relevant actions or states that don't allow the goals to be reached. Goal states manifest as particular nodes that match the problem solution objective.

Search graphs have numerous design advantages. They draw each state once and never omit any state based on the count of paths that reach that state. This ensures that the algorithm does not revisit states that it has visited before. Algorithms that use graphs can use cycle detection and avoid loops and also recognize if multiple paths reach the same state.

The representation efficiency of the graph lies in the determination of whether two paths reach the same state. This requires an adequate state comparison mechanism that will identify states that are the same regardless of how they came into existence.

5.4.2   Search Trees: Generation and Analysis

A search tree represents the space-exploring process and not the space itself [1]. The nodes of a tree represent states of the space of the search and the state will be represented multiple times within the tree in different branches. The tree paths leading to any node represent a unique set of actions.

The root of the tree is the initial state. The nodes branch with respect to the relevant actions in the state they inhabit. The node n with state s has children consisting of each state s' such that $s' = T(s, a)$ with an action $a \in A$. The tree branches further with the algorithm attempting sequences of actions.

In a search tree, the nodes contain two kinds of important information:

1. State that describes the node that matches the representation;
2. Record of the nodes that came leading to the node from the root.

Path information typically includes the order of the steps and possibly the total path cost. This will be utilized by the algorithm in the construction of the solution path once a solution state has been located. Path-dependent evaluations that not only include the state reached but also how it has been reached are also provided.

The tree and the problem it represents has a less straightforward correlation with respect to the search graphs. The tree represents paths of possibilities and not the real shape of the search space. This results in trees being an excellent representation of the execution of a search algorithm that tests sequences of activities without necessarily knowing how they will be fruitful initially.

Search tree efficiency has to do with the potentially infinite tree size. The tree could be infinite if the state space of the problem has cycles. Even if the tree does not have cycles, it has an exponential factor of depth with respect to the branching factor. This exponential factor prevents exhaustive tree traversal of non-trivial problems in general.

To manage this complexity, practical search algorithms must be selective about which branches to explore. Various strategies such as depth limits, heuristic evaluation, and pruning techniques help focus exploration on promising areas of the tree. These approaches aim to find solutions without exhaustive tree expansion.

5.4.3   Differences Between Graphs and Trees

The major difference between search trees and search graphs lies in how they handle multiple paths that arrive at the same state [1]. This has profound implications with respect to algorithm design and execution.

Search graphs illustrate a state once and once only irrespective of the num-ber of distinct paths that arrive at a state. This representation saves memory space in problem instances with lots of similar paths that arrive at the same states. Algorithms based on the application of a graph minimize redundancy of explor-ation if a found state has been visited before. This aspect comes in handy with problem instances with lots of similar paths and cycles.

Search trees instead record each individual path separately although paths may reach the same state. This constitutes a potentially larger yet simpler-to-explore tree. Tree techniques never need to ask if the states have been visited and avoid computation at the price of memory usage. Tree representation also maintains the whole set of actions leading to each state with ease.

The choice between graph and tree representations affects several aspects of search algorithm implementation (Fig. 5.7):

1. Memory consumption: Trees use more memory since each state will be stored multiple times within a tree compared to a graph that saves the state once.
2. Handling cycles: Tree algorithms need not necessarily identify cycles explicitly in order not to get stuck in infinite loops, whereas graph algorithms need to do this explicitly.
3. Optimality guarantees: Shortest-path finding problems make it easier for the graph methods to assure optimality since they expand the shortest path leading to each state at a time.
4. Implementational simplicity: Tree algorithms are simpler to implement because they don't need state recognition mechanisms that identify visited states. Tree

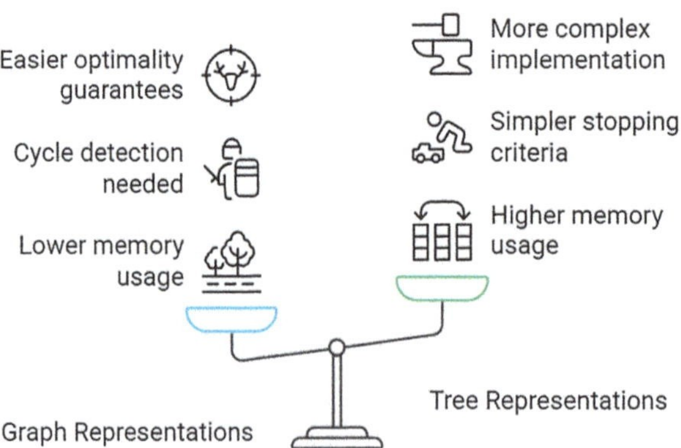

Fig. 5.7 Compare search algorithm tradeoffs

algorithms require good state storage and state retrieval facilities in the data structures.

Generally speaking, actual implementations of the search algorithms use hybrid techniques that borrow elements of each representation. Tree-search and graph-search implementations of the familiar breadth-first and depth-first algorithm variants use an active record of visited states and avoid them explicitly revisiting them, effectively overlaying a graph representation of the state space on the algorithmic state transition and exploration processes. Tree-search variants might prioritize efficient exploration with minimal record-keeping of visited states.

5.5   Practical Examples and Applications

Abstract problem formulation and notions of the search space are understood better with real-life illustrations [1]. The section considers three typical issues that portray different problem formulation notions, state representation, and how a good search strategy needs to be identified.

5.5.1   The 8-Puzzle Problem

There are eight tiles with labels and a 3×3 grid with a blank space in the 8-puzzle. The task is to get the tiles from an initial state to a final state by sliding the tiles into the blank space. This seemingly simple puzzle constitutes an excellent problem definition and space analysis study (Fig. 5.8).

A complete definition of the 8-puzzle as a search problem includes specifying each component of the quintuple $P = (S, A, T, G, C)$.

State space S of the 8-puzzle encompasses any combination of the eight tiles and the blank space within the 3×3 square. One state may be a 3×3 array of the contents of each square. The state also may be a string or array of nine symbols representing the tile 1–8 or the blank space (often 0) in the corresponding position.

For example, the goal state might be represented as:

$$
s_{\text{goal}} = \begin{bmatrix} 1 & 2 & 3 \\ 8 & 0 & 4 \\ 7 & 6 & 5 \end{bmatrix}.
$$

The set of actions A includes four movements of a tile into the vacant space: up, down, left, and right. Yet not every action is feasible in any state. The feasible actions vary with the position of the vacant space. Four movements are feasible if the vacant space is in the center position. Two movements are feasible if the vacant

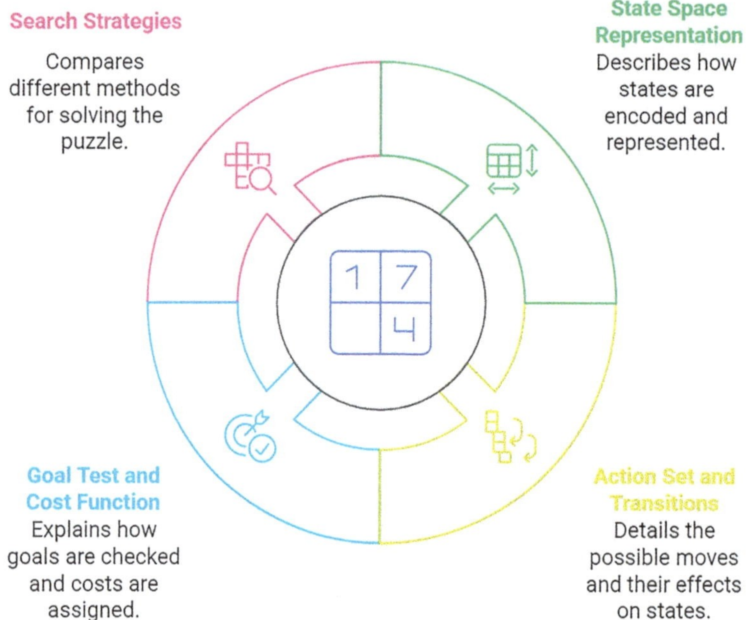

Fig. 5.8 8-Puzzle Problem Analysis

space is in a corner position. Three movements are feasible if the vacant space is an edge position and not a corner position.

Transition function T indicates how an action will transition a state to a state that is different. Let us use the example of the blank space at position (i, j) and the move tile from the left action. The state that will be the result will be the blank space at position $(i, j - 1)$ and the tile that was at position $(i, j - 1)$ will be at position (i, j).

Goal test G checks whether the state at the current time is the target state. This is typically carried out through a comparison of the state at the current time with the target state.

Cost function C of the 8-puzzle typically charges a constant cost of 1 per move so that the solution path cost will be the number of moves performed.

There are a few notable properties of the 8-puzzle space of search. Its space has a bounded size of $9! = 362, 880$ tile configurations. However, half of the states are inaccessible from any state due to the constraint of parity. The branching factor also varies between 2 and 4 depending on the position of the blank space and has an average of 3.

The maximum solution depth for the 8-puzzle is 31 moves. This means that any solvable configuration can be resolved in at most 31 moves. The average solution depth for randomly generated puzzles is around 22 moves. These properties make the 8-puzzle challenging enough to illustrate search algorithms while remaining tractable for analysis.

Different methods of search can be used with the 8-puzzle. Uninformed methods such as breadth-first search will find the optimal solution with a promise of completeness and a penalty of potentially examining a large number of states. Informed methods such as the use of the Manhattan distance heuristic with the A* algorithm will find the optimal solution with a large improvement in efficiency.

5.5.2 The Traveling Salesman Problem

The Traveling Salesman Problem (TSP) forms a special class of the search problem. Here the salesman needs to travel between a set of cities and return home with the least distance traveled. The problem of the TSP captures the essence of the difficulty of the combinatorial optimization and how the problem formulation affects the solution strategy.

There are numerous ways the state space S of TSP can be modeled. One of them states that a state will be a partial tour of visited cities. Here a state can be an ordered set of visited cities and the distance traveled till now.

We can define a state in an n cities problem as $s = (c_1, c_2, \ldots, c_k)$ with the starting city c_1, the current city c_k, and the remaining cities visited in that specific order. The state might also track the set of the remaining cities that need to be visited.

Set of actions A constitutes the choice of the next-to-visit city. The set of accessible actions at any state constitutes a move to any of the remaining-to-be-visited cities. The set of accessible actions reduces with each move until the final move returns the customer to the starting city.

Transition function T extends the current tour with the inclusion of the additional city. Assuming the current state is $s = (c_1, c_2, \ldots, c_k)$ and the move includes the travel of the city c_{k+1}, the state that will be reached will be $s' = (c_1, c_2, \ldots, c_k, c_{k+1})$. The transition also increases the travel distance with the distance between c_k and c_{k+1}.

Goal state G guarantees that the tour has visited each of the cities once and that the tour returns home. Under the state representation that we've been using, a goal state would be of the form $s_{goal} = (c_1, c_2, \ldots, c_n, c_1)$, with each of the cities appearing once except that the starting and ending position of the state also includes the occurrence of c_1.

Cost function C adds the distance between two neighboring cities in the tour. The entire cost of a complete tour $(c_1, c_2, \ldots, c_n, c_1)$ will be $\sum_{i=1}^{n} d(c_i, c_{i+1})$, with the convention that $c_{n+1} = c_1$.

The problem space of the TSP factorially increases with the number of cities. There are $(n-1)!$ tours if we include the starting city and the cyclic nature of tours. This exponential growth does not allow exhaustive search with anything except the smallest instances. The branch factor reduces with the construction of the tour, initially at $n-1$, and reducing at each step.

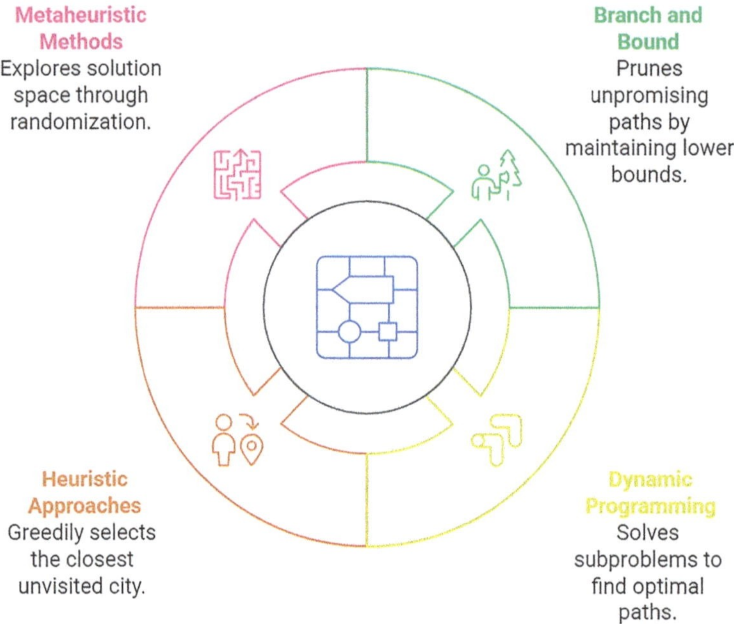

Fig. 5.9 Traveling Salesman Problem Strategies

Owing to the problem of the combinatorial complexity of TSP, it is normally addressed with specialized techniques instead of general search methods. They include (Fig. 5.9):

1. Branch and bound techniques that prune unpromising paths based on lower bounds of the tours' cost.
2. Programming techniques that recursively solve the subproblems of finding the best paths through the subset of cities.
3. Algorithms such as the Nearest-Neighbor algorithm that greedily selects the shortest distance yet-to-be-visited city at each move.
4. Metaheuristics such as the application of genetic algorithmic techniques or annealing that move through the solution space with randomized controls.

The TSP demonstrates how problem structure directs the solution strategy choice. Its combinatorial character prevents uninformed search and instead favors problem-specific exploitation strategies.

5.5.3 Route Finding and Navigation Problems

Route finding and navigation issues illustrate how the methods of search are relevant to real and applied issues. They include the determination of the best paths within physical space and abstractions of space based on distance, travel time, cost, and safety factors. Navigation issues appear in GPS technology, robots, network routing, and numerous other contexts.

Typically, the state representation in a route finding problem will be the position and possibly additional attributes such as direction of travel, remaining fuel, or the time of the day. The state may be a representation of a road navigation problem such as $s = (l, t)$, where l representing the position of the place we're at the moment (road segment or intersection) and t representing the distance traveled or the time of the day.

The set of movements involves travel between the current position and neighboring positions along accessible paths. Actions within a network of roads equate to travel on specific roads to neighboring crossroads. The set of accessible movements depends on the position and physical constraint such as a one-way street or a restricted vehicle.

Transition computes the state after traversing a particular route. This will update the position and possibly update any of the state variables such as the time or the fuel remaining. The transition includes real-world limitations such as travel times that will be dependent on the state of the traffic and the time of day.

Goal testing checks whether the current position matches the target position. The goal may also include additional specifications in a particular situation, such as arrival within a specific period of time and with a specific reserve of fuel.

Cost normally encompasses measuring travel distance, travel time, or a combination of variables. Realistic routing often involves multiple criterion optimization with a tradeoff among variables such as distance, travel time, fuel consumption, toll charges, and route preference.

Graph methods inherently frame navigation tasks. Nodes relate to positions and edges relate to paths. Weights of the edges stand for travel expense (distance, duration, etc.). This formulation allows the use of generic graph traversal methods such as Dijkstra's algorithm or A*.

Development of heuristics constitutes an integral part of good route finding. The typical heuristics comprise (Fig. 5.10):

1. Euclidean (straight line) distance to the target, providing a lower estimate of the true travel distance.
2. Manhattan distance in grid environments that sums the distance horizontally and the distance vertically.
3. Landmark-based heuristics that use precomputed distances to landmark points to improve estimation accuracy.

Real world navigation involves a multitude of practical issues beyond standard search formulation (Fig. 5.11):

1. Environment that dynamically varies with travel times that change with traffic, weather, and the like. This may call for real-time updating of the graph representation or modification of the search strategy.
2. Hierarchical techniques that design at multiple scales of detail, starting with large highways or zones and working downward toward detailed street-level paths.

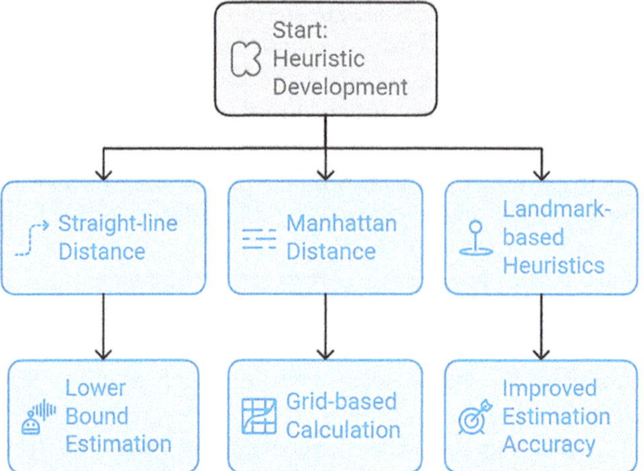

Fig. 5.10 Heuristic Development in Route Finding

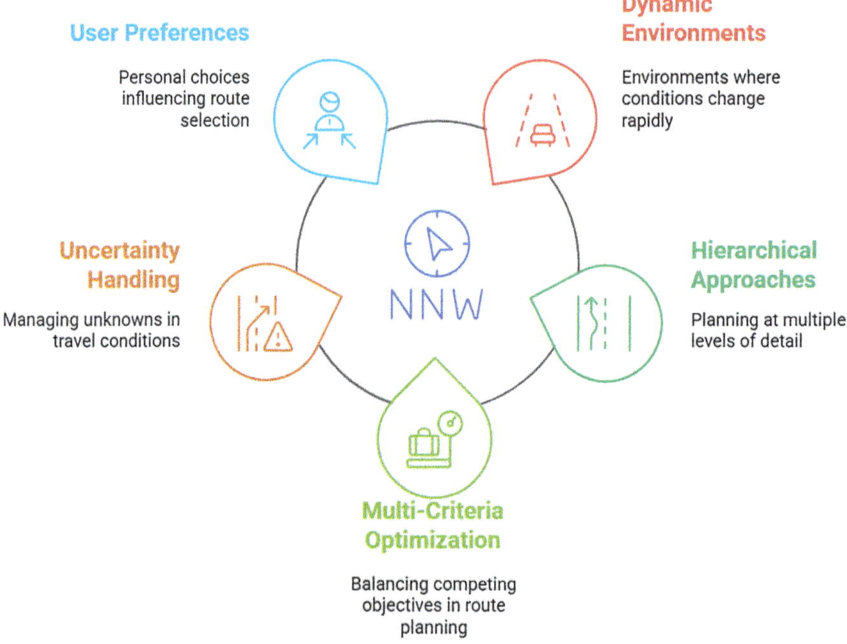

Fig. 5.11 Considerations for Effective Navigation

3. Multi-object optimization that balances tradeoffs including minimizing distance and the absence of tolls and congestion.
4. Management of uncertainty in the situation of travel times that aren't exactly known and require treatments based on probability.
5. Preferential routing based on non-monetary reasons such as route preference based on scenery and avoiding toll roads.

These real-world examples demonstrate how abstractions of the search concepts manifest in real life. The problem domain of each emphasizes distinct problem formulation and the choice of search strategy. The 8-puzzle emphasizes state representation and heuristic search in a constraint environment. The TSP emphasizes the combinatorial optimization problem and the use of specialized methods. The navigation problem emphasizes how the techniques of search fit into real-world practical limitations and goals.

By examining such examples, we get a good notion of how to generate new instances of the problem and how to select appropriate solution techniques. The techniques demonstrated here extend to a wide array of tasks throughout artificial intelligence and computational problem-solving.

5.6   Practice and Exercises

This section provides hands-on activities to reinforce understanding of problem formulation and search spaces. The exercises are designed to develop practical skills in analyzing and solving AI problems through systematic application of the concepts covered in this chapter. Working through these exercises will strengthen your ability to formulate problems, analyze search spaces, and select appropriate solution strategies.

5.6.1   Self-Assessment Questions

These questions are designed to test your understanding of key concepts presented in the chapter. For each question, select the answer you believe is correct. This exercise will help solidify your knowledge and identify areas that may need additional review.

1. What primarily distinguishes a well-formulated problem in AI?

    (a) Running time of the solution algorithm
    (b) Clear definition of states, actions, transitions, goals, and costs
    (c) Number of variables in the state representation
    (d) Implementation language used for the solution

2. In a formal problem definition $P = (S, A, T, G, C)$, what does the transition function T represent?

   (a) The time required to solve the problem
   (b) How states change when actions are applied
   (c) The test that determines if a state is a goal
   (d) The topology of the search space

3. Which property of a search space most directly affects the memory requirements of breadth-first search?

   (a) Branching factor
   (b) Solution depth
   (c) Total number of states
   (d) Number of goal states

4. When using factorial representation for states, what happens to the size of the state space as the number of variables increases?

   (a) It increases linearly
   (b) It increases exponentially
   (c) It remains constant
   (d) It decreases proportionally

5. What is the main difference between a search graph and a search tree?

   (a) Search trees can only be used for depth-first search
   (b) Search graphs cannot represent goal states
   (c) In a search tree, the same state may appear multiple times in different branches
   (d) Search graphs cannot handle cycles in the state space

5.6.2   *Conceptual Exercise: Vacuum World Problem*

Consider the Vacuum World problem, a simple environment consisting of two rooms (A and B) side by side. In each room, there may be dirt that needs to be cleaned. A vacuum cleaner agent can move between rooms and can clean dirt from a room.

Task
- Provide a complete formulation of this problem as a search problem:
- Define the state space S. How would you represent each state? How many possible states are there?
- Specify the action set A. What actions can the agent perform?
- Describe the transition function T. How do the actions change the state of the environment?
- Formulate the goal test G. What states satisfy the objective of the problem?

- Define an appropriate cost function C. What costs would you assign to different actions?
- Draw a state space graph for this problem, showing all states and transitions.
- Analyze the properties of the search space:
- What is the branching factor?
- What is the maximum solution depth from the worst initial state?
- Would you recommend breadth-first or depth-first search for this problem? Explain your reasoning.

5.7 Conclusions and Future Directions

Problem formulation and search space analysis form the foundation of effective AI problem-solving. This chapter has explored systematic approaches to transform real-world challenges into well-structured computational problems. These concepts enable more efficient algorithms and better solutions across diverse domains.

The formal problem definition framework using the quintuple $P = (S, A, T, G, C)$ provides a powerful approach to representing challenges. This framework enables precise definition of states, actions, transitions, goals, and costs. Such formalization transforms vague descriptions into algorithmically solvable representations. This approach identifies essential problem components while eliminating unnecessary complexity.

State representation significantly impacts problem-solving efficiency. The choice between atomic and factorial representations involves trade-offs between simplicity and expressiveness. Well-designed representations highlight relevant features while minimizing state space size. They also facilitate heuristic development and efficient state transitions. The representation choice often determines whether a problem can be solved with reasonable computational resources.

Search space analysis reveals fundamental properties that influence algorithm selection. Understanding dimensionality, branching factor, and solution depth helps predict computational requirements. This analysis identifies challenges like plateaus, local optima, or disconnected regions. Examining these properties helps select appropriate algorithms and design effective heuristics.

Search graphs and trees provide complementary representations for algorithm design. Graphs efficiently represent state spaces by eliminating redundancy. Trees explicitly capture exploration paths. This distinction affects memory usage, cycle handling, and optimality guarantees. Many practical algorithms combine both approaches to balance efficiency and effectiveness.

The examples in this chapter demonstrate these concepts in real-world applications. The 8-puzzle, traveling salesman problem, and navigation challenges highlight different aspects of problem formulation. These examples provide templates adaptable to similar problems across various domains. They illustrate how problem characteristics guide solution strategy selection.

References

1. Norvig, P.: Artificial intelligence: a modern approach. Global edition. Pearson, Boston (2021).
2. Nilsson, N.J.: The Quest for Artificial Intelligence. Cambridge University Press (2009).
3. Back, T., Fogel, D.B., Michalewicz, Z. eds: Handbook of Evolutionary Computation. Oxford University Press, Bristol; Philadelphia: New York (1997).
4. Castillo, O., Melin, P., Pedrycz, W. eds: Hybrid Intelligent Systems: Analysis and Design. Springer, Berlin Heidelberg (2007). https://doi.org/10.1007/978-3-540-37421-3.

Chapter 6
Uninformed Search Algorithms

Abstract This chapter presents uninformed search algorithms, a fundamental class of problem-solving techniques in artificial intelligence. These algorithms operate without domain-specific knowledge. The discussion builds on concepts of state spaces and problem representation. We explore how these algorithms systematically traverse search spaces to find solutions. The chapter provides both theoretical foundations and practical implementations. Examples and Python code demonstrate real-world applications.

6.1 Introduction to Uninformed Search

Uninformed search algorithms, also known as blind search or brute-force search methods, represent the foundational approach to problem-solving in artificial intelligence [1, 2]. These algorithms work systematically through a search space without using any problem-specific knowledge beyond the basic problem definition [1, 3].

6.1.1 Basic Concepts and Problem Representation

The fundamental idea of uninformed search is remarkably straightforward. We begin with a problem defined by its initial state, possible actions, and goal conditions. The search process systematically explores different states until it finds a solution or exhausts all possibilities [1, 4].

Formally, we can define a search problem as a tuple:

$$P = \left(S, A, T, s_0, G \right),$$

where:

- S is the set of possible states;
- A is the set of available actions;

O. Kuznetsov, *Intelligent Systems: From Theory to Applications*, Cognitive Technologies, https://doi.org/10.1007/978-3-032-00044-6_6

- $T : S \times A \to S$ is the transition function;
- $s_0 \in S$ is the initial state;
- $G \subseteq S$ is the set of goal states.

This formal representation provides a framework for understanding any search problem [5]. For example, in a simple navigation problem, S might represent different locations, A the possible movements, and G the destination.

6.1.2   Historical Context

The development of uninformed search algorithms traces back to the early days of artificial intelligence. In the 1950s, researchers like Newell, Simon, and Moore introduced these fundamental concepts while working on the General Problem Solver [6]. These early efforts established the groundwork for systematic problem-solving in AI [4, 7].

The evolution of these algorithms closely parallels the development of computer science itself. Early implementations were limited by computational resources, but they established crucial principles that remain relevant today.

6.1.3   Importance in AI Problem Solving

Uninformed search algorithms serve several critical roles in artificial intelligence [8]:

- Foundational Understanding. These algorithms provide the basic framework for understanding how automated problem-solving works. They demonstrate essential concepts like state spaces, transitions, and solution paths.
- Benchmark Methods. They serve as baseline approaches against which more sophisticated algorithms can be compared. This comparison helps in understanding the value of additional problem-specific knowledge.
- Practical Applications. Despite their simplicity, uninformed search methods remain useful in many real-world scenarios. They are particularly valuable when domain-specific knowledge is unavailable or unreliable.

6.1.4   Search Space Characteristics

Understanding search spaces is crucial for applying these algorithms effectively. A search space consists of [3, 7]:

- States: Representations of possible situations;
- Actions: Ways to move between states;

- Path Cost: The cost associated with taking actions;
- Branching Factor: The average number of actions available per state.

The structure of this space significantly impacts algorithm performance. For instance, a search space with many redundant paths might require additional mechanisms to avoid cycles.

6.1.5 Basic Search Process

The general search process follows a simple pattern (Fig. 6.1):

- Start at the initial state;
- Maintain a frontier of states to explore;
- Repeatedly select and expand states from the frontier;
- Check if each new state is a goal state;
- Continue until finding a solution or exhausting all possibilities.

Figure 6.1 illustrates the fundamental steps of the search process. It shows how we start from an initial state, explore possible actions, and check for goal states. This simple flow helps us understand the basic mechanics of all search algorithms.

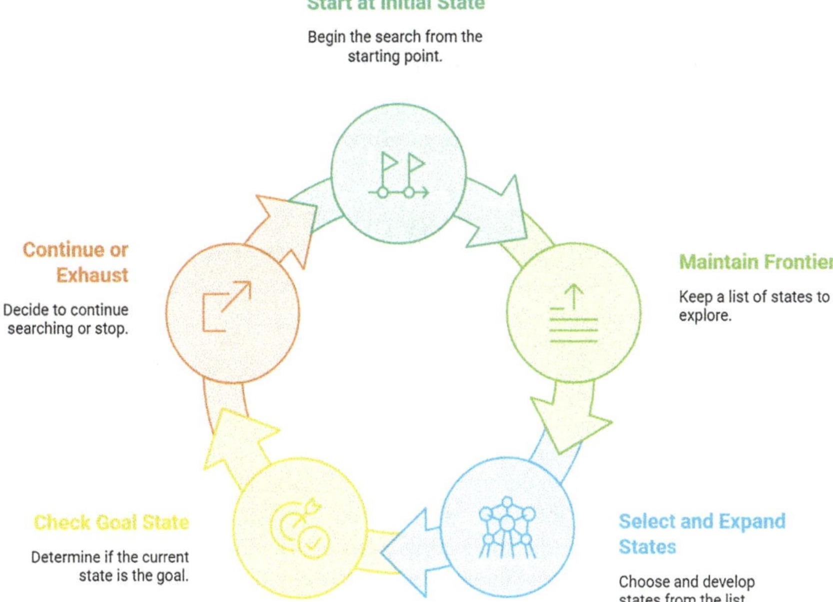

Fig. 6.1 The basic search process

Different uninformed search strategies vary primarily in how they manage this frontier of unexplored states. This variation leads to different behaviors and properties in terms of completeness, optimality, and efficiency.

The key advantage of uninformed search lies in its generality and simplicity. These algorithms require no additional knowledge beyond the problem definition itself. This makes them robust and widely applicable, though often at the cost of efficiency compared to more informed approaches [1, 2].

Understanding these fundamental concepts provides the foundation for exploring specific uninformed search algorithms in the following sections. These algorithms build upon these basic principles to provide different trade-offs in terms of memory usage, completeness, and optimality.

This introduction sets the stage for a detailed examination of specific uninformed search algorithms, their implementations, and their practical applications. The following sections will explore these algorithms in detail, starting with the most basic approaches and progressing to more sophisticated variants.

6.2   Theoretical Foundations

The foundation of uninformed search algorithms rests on several key theoretical concepts. These concepts provide the framework for understanding how search algorithms work and how they can be implemented effectively.

6.2.1   State Space Representation

The state space represents the universe of all possible states in a problem. We can formally define it as a graph [2, 3]:

$$G = (V,E),$$

where:

- V is the set of vertices, each representing a state;
- E is the set of edges, representing possible transitions between states.

For any practical problem, the state space must be represented efficiently. For example, in a pathfinding problem, each state might represent a position in a grid. The transitions between states would represent possible movements.

The dimensionality of the state space directly impacts the complexity of the search problem. In real-world applications, this space can be extremely large or even infinite. This makes efficient state representation crucial for practical implementations.

6.2.2 Search Trees and Graphs

A search tree represents the actual exploration of the state space. Each node in the search tree represents a specific state and contains [8]:

- State: *state(n)*—The current configuration;
- Parent: *parent(n)*—Reference to the preceding node;
- Action: *action(n)*—The action that led to this state;
- Path Cost: $g(n)$—The cost from the initial state to node n.

The relationship between state space graphs and search trees is fundamental. While the state space might contain cycles, the search tree explicitly shows the path taken to reach each state.

6.2.3 Path Costs and Optimality

Path cost represents the cumulative cost of reaching a state. For a path consisting of actions $[a_1, a_2, \ldots, a_k]$, the total cost is [8]:

$$g(n) = \sum_{i=1}^{k} cost(a_i),$$

The concept of optimality in search relates to finding paths with minimal cost. A solution is optimal if no other path to the goal has a lower total cost.

Key properties related to path costs include:

- Path cost monotonicity;
- Goal path optimality;
- Cost consistency between states.

6.2.4 Problem Formulation Components

A well-formed search problem consists of five essential components [8]:

- Initial State (s_0) The starting point of the search.
- Actions Function $A(s)$ returns the set of actions possible in state s.
- Transition Model $T(s, a)$ returns the state resulting from action a in state s.
- Goal Test A function that determines whether a given state is a goal state.
- Path Cost Function $c(s, a, s')$ provides the cost of taking action a from state s to reach state s'.

### 6.2.5	Search Strategy Classification

Search strategies can be evaluated based on four key criteria [4, 7]:

- Completeness. Whether the algorithm guarantees finding a solution if one exists.
- Time Complexity. The maximum number of nodes that must be examined.
- Space Complexity. The maximum number of nodes that must be stored in memory.
- Optimality. Whether the algorithm guarantees finding the optimal solution.

These metrics help us understand the trade-offs between different search strategies. For instance, breadth-first search guarantees optimality for uniform-cost actions but requires significant memory. Depth-first search uses minimal memory but may not find optimal solutions.

Understanding these theoretical foundations is crucial for:

- Implementing search algorithms effectively;
- Choosing appropriate strategies for specific problems;
- Analyzing algorithm performance;
- Identifying potential optimizations.

The following sections will build upon these foundations to examine specific uninformed search algorithms and their practical implementations. Each algorithm makes different trade-offs among these fundamental aspects, leading to distinct characteristics and applications.

This theoretical framework provides the basis for understanding why different search strategies behave as they do and when each might be most appropriate. It also helps in recognizing the limitations and potential improvements of basic search algorithms.

## 6.3	Basic Search Algorithms

Uninformed search algorithms provide systematic approaches to exploring state spaces without using domain-specific knowledge. Each algorithm offers different trade-offs between memory usage, completeness, and optimality. Understanding these fundamental algorithms provides the foundation for more advanced search techniques.

### 6.3.1	Breadth-First Search (BFS)

Breadth-First Search represents one of the most fundamental and widely used search algorithms [4, 9]. It systematically explores the search space level by level, ensuring that all nodes at the current depth are examined before moving to the next level.

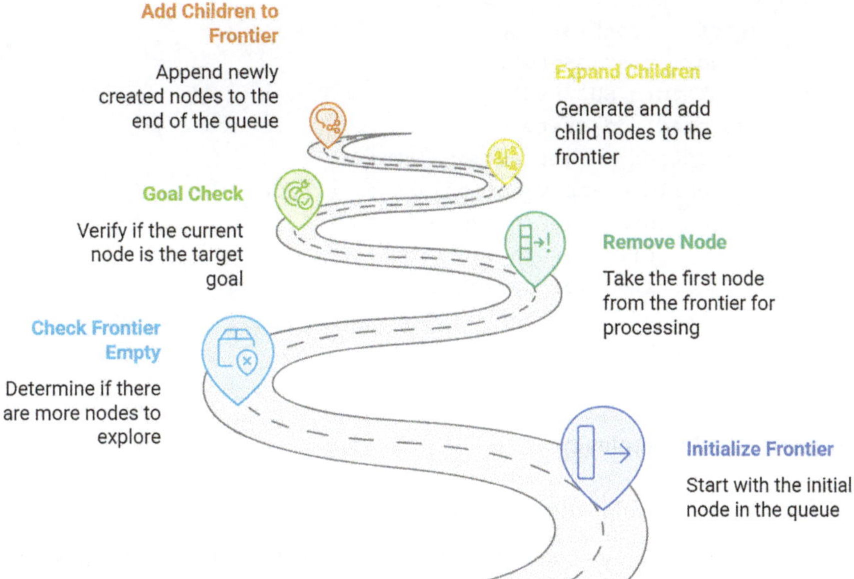

Fig. 6.2 Breadth-First Search algorithm flow

Algorithm Description

BFS operates by maintaining a queue of unexplored nodes, called the frontier. The algorithm follows a simple but effective process (Fig. 6.2):

1. Initialize the frontier with the start node.
2. While the frontier is not empty:

   - Remove the first node from the frontier;
   - If this node is the goal, return success;
   - Otherwise, expand all its children;
   - Add the children to the end of the frontier.

Figure 6.2 demonstrates how Breadth-First Search (BFS) works step-by-step. It clearly shows the queue-based approach where all nodes at one level are explored before moving to the next level. This visual representation makes it easy to grasp BFS's systematic exploration strategy.

Here is the formal algorithm implementation in pseudocode:

```
function BFS(initial_state, goal_test):
    frontier = Queue()
    frontier.add(initial_state)
    explored = Set()
```

```
while not frontier.is_empty():
    state = frontier.pop()
    if goal_test(state):
        return SUCCESS
    explored.add(state)
    for action in get_actions(state):
        child = apply_action(state, action)
        if child not in explored and
            child not in frontier:
              frontier.add(child)


return FAILURE
```

Properties and Complexity

BFS has several important theoretical properties (Fig. 6.3):

1. Completeness: BFS is complete when the branching factor is finite. It will always find a solution if one exists.
2. Time Complexity: $O(b^d)$, where:
   - b is the branching factor (average number of successors);
   - d is the depth of the shallowest solution.

3. Space Complexity: $O(b^d)$, as it must store all nodes at the current and next levels.
4. Optimality: BFS is optimal for unit-cost actions, meaning it finds the shortest path in terms of the number of actions.

Figure 6.3 summarizes the key advantages and disadvantages of BFS. It highlights BFS's ability to find the shortest path while also pointing out its high memory requirements:

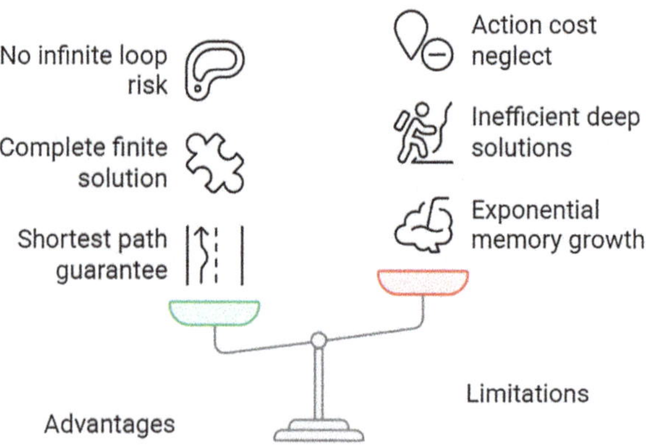

Fig. 6.3 Strengths and limitations of Breadth-First Search

Advantages:
- Guaranteed to find the shortest path in terms of the number of steps;
- Complete solution for finite search spaces;
- Systematic exploration without risk of infinite loops;
- Simple implementation using a queue data structure.

Limitations:
- Memory requirements grow exponentially with depth;
- May be inefficient for deep solutions;
- Does not consider action costs (when they are non-uniform);
- Cannot handle infinite search spaces effectively.

Understanding these properties helps us evaluate when BFS is appropriate for a given problem.

Implementation Considerations

The practical implementation of BFS requires careful attention to several details:

1. Queue Implementation:

```python
from collections import deque
class Queue:
    def __init__(self):
        self.items = deque()

    def add(self, item):
        self.items.append(item)

    def pop(self):
        return self.items.popleft()
```

2. State Management:

```python
def get_successors(state):
    successors = []
    for action in get_valid_actions(state):
        new_state = apply_action(state, action)
        successors.append(new_state)
    return successors
```

BFS provides a fundamental template for understanding search algorithms. Its systematic approach makes it particularly useful for educational purposes and serves as a building block for more sophisticated search strategies. In practice,

BFS is most effective for problems with shallow solutions or when memory constraints are not a significant concern.

The next sections will explore other uninformed search algorithms that make different trade-offs between memory usage, completeness, and optimality. Understanding these alternatives helps in selecting the most appropriate algorithm for specific problem characteristics.

6.3.2   Depth-First Search (DFS)

Depth-First Search represents a fundamentally different approach to exploring search spaces [4, 10]. Instead of examining states level by level, DFS aggressively pursues a single path as deeply as possible before backtracking to explore alternatives.

Algorithm Description

DFS explores the search space by always expanding the deepest unexpanded node. The algorithm maintains a stack of nodes representing the current exploration path. The basic process follows these steps (Fig. 6.4):

1. Initialize the frontier with the start node
2. While the frontier is not empty:

   - Remove the last node from the frontier;
   - If this node is the goal, return success;

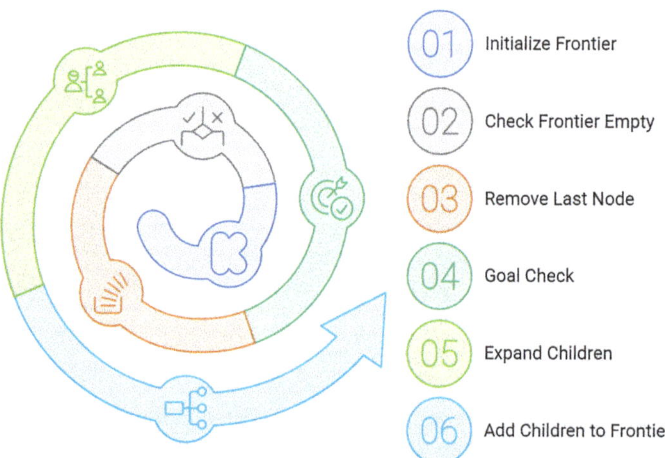

Fig. 6.4 Depth-First Search algorithm flow

- Otherwise, expand all its children;
- Add the children to the top of the frontier.

Figure 6.4 explains the Depth-First Search (DFS) process visually. It shows how DFS explores as far as possible along each branch before backtracking. This clear depiction helps learners understand DFS's unique exploration pattern compared to BFS.

Here is the formal algorithm implementation in pseudocode:

```
function DFS(initial_state, goal_test):
    frontier = Stack()
    frontier.push(initial_state)
    explored = Set()

    while not frontier.is_empty():
        state = frontier.pop()
        if goal_test(state):
            return SUCCESS
        explored.add(state)
        for action in get_actions(state):
            child = apply_action(state, action)
            if child not in explored and
                child not in frontier:
                frontier.push(child)


    return FAILURE
```

Properties and Complexity

DFS exhibits several distinctive theoretical properties:

1. Completeness: DFS is complete only for finite search spaces. It may not terminate in infinite spaces.
2. Time Complexity: $O(b^m)$, where:
   - b is the branching factor;
   - m is the maximum depth of the search space;

3. Space Complexity: $O(bm)$, where:
   - Only needs to store nodes on the current path;
   - Much more space-efficient than BFS;

4. Optimality: DFS does not guarantee finding the optimal (shortest) path to the goal.

Implementation Details

The practical implementation requires careful attention to the stack data structure:

```python
class Stack:
    def __init__(self):
        self.items = []

    def push(self, item):
        self.items.append(item)

    def pop(self):
        return self.items.pop()

    def is_empty(self):
        return len(self.items) == 0
```

The node expansion function remains similar to BFS, but the order of exploration differs significantly:

```python
def expand_node(state):
    successors = []
    for action in get_valid_actions(state):
        new_state = apply_action(state, action)
        successors.append(new_state)
    # Note: Often process successors in reverse order
    return reversed(successors)
```

Figure 6.5 outlines the main advantages and limitations of DFS. It emphasizes DFS's memory efficiency while also noting its potential to get stuck in infinite paths:

Advantages:

1. Memory Efficiency

   - Requires storage only proportional to search depth;
   - Particularly useful for deep search spaces;

2. Implementation Simplicity

   - Natural recursive implementation possible;
   - Simple stack-based iterative version;

3. Problem-Solving Characteristics

   - Can find solutions quickly in deep search spaces;
   - Useful for problems where solutions are likely to be deep;

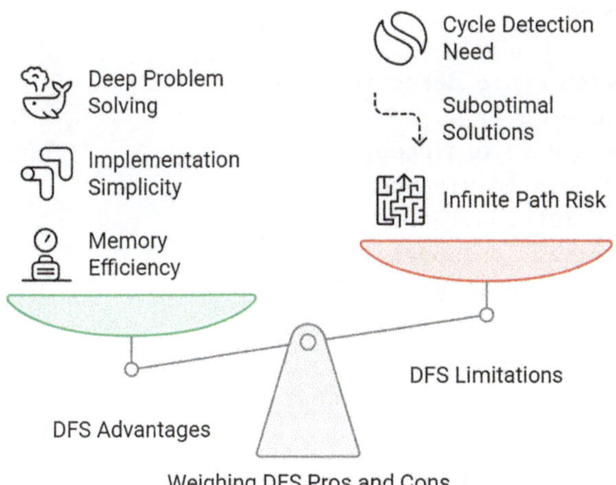

Fig. 6.5 Strengths and limitations of Depth-First Search

Limitations:

1. Search Behavior

    - May get trapped in infinite paths without cycle detection;
    - Does not guarantee shortest solutions;

2. Solution Quality

    - Can find extremely deep, suboptimal solutions;
    - No relationship between search depth and solution cost;

3. Practical Considerations

    - Requires cycle detection in graphs;
    - Stack overflow possible in recursive implementations;

This summary helps us understand the trade-offs involved in using DFS for different types of problems.

DFS finds particular utility in:

- Memory-constrained environments;
- Problems with solutions deep in the search space;
- Scenarios where finding any solution quickly is more important than finding the optimal solution.

Implementation Considerations

When implementing DFS, several practical considerations become important:

1. Cycle Detection:

```python
def dfs_with_cycle_detection(start):
    visited = set()
    def recursive_dfs(node):
        if node in visited:
            return
        visited.add(node)
        for child in get_children(node):
            recursive_dfs(child)
```

2. Depth Limiting:

```python
def depth_limited_dfs(node, depth_limit):
    if depth_limit == 0:
        return
    for child in get_children(node):
        depth_limited_dfs(child, depth_limit - 1)
```

DFS provides a complementary approach to BFS, making different trade-offs between memory usage and solution optimality. Understanding these trade-offs is crucial for selecting the appropriate search strategy for specific problem domains. The next section will explore Uniform-Cost Search, which introduces consideration of action costs into the search process.

6.3.3 Uniform-Cost Search (UCS)

Uniform-Cost Search extends basic search algorithms by considering action costs [11]. UCS always expands the node with the lowest path cost, ensuring optimal solutions in scenarios with varying action costs.

Algorithm Description

UCS maintains a priority queue of nodes ordered by their path cost from the start state. The basic process follows these steps:

1. Initialize the frontier with the start node (cost 0)
2. While the frontier is not empty:

   - Remove the lowest-cost node from the frontier;
   - If this node is the goal, return success;
   - Otherwise, expand all its children;
   - Add children to the frontier with their cumulative costs.

Fig. 6.6 Uniform-Cost
Search algorithm flow

Add Children to Frontier

Add the children nodes to the frontier
with their cumulative costs.

Expand Children Nodes

If not the goal, expand the node's
children to explore further.

Check for Goal

Determine if the removed node is the
goal node.

Remove Lowest-Cost Node

Select and remove the node with the
lowest path cost from the frontier.

Initialize Frontier

Begin with the start node, setting its
cost to zero.

Figure 6.6 outlines the Uniform-Cost Search (UCS) process.

It demonstrates how UCS uses a priority queue to always expand the lowest-cost node first. This visualization makes it easier to understand how UCS incorporates path costs into its decision-making.

Here is the formal algorithm implementation in pseudocode:

```
function UCS(initial_state, goal_test):
    frontier = PriorityQueue()
    frontier.add(initial_state, 0)
    explored = Set()

    while not frontier.is_empty():
        state, cost = frontier.pop()
        if goal_test(state):
            return SUCCESS
        explored.add(state)
        for action in get_actions(state):
            child = apply_action(state, action)
            child_cost = cost + action_cost(state, action)
```

```
        if child not in explored and
            child not in frontier:
              frontier.add(child, child_cost)
          elif child in frontier with higher cost:
              frontier.update(child, child_cost)


  return FAILURE
```

Properties and Complexity

UCS has several important theoretical properties:

1. Completeness: Complete when all action costs are greater than some small positive constant ε.
2. Time Complexity: $O\left(b^{1+C^*/\varepsilon}\right)$, where:
   - b is the branching factor;
   - C^* is the cost of the optimal solution;
   - ε is the minimum action cost;
3. Space Complexity: $O\left(b^{1+C^*/\varepsilon}\right)$.

   - Must store all nodes with cost less than or equal to optimal solution;

4. Optimality: Guaranteed to find the lowest-cost path to the goal.

Implementation Details

The priority queue implementation is crucial for UCS:

```python
import heapq
class PriorityQueue:
    def __init__(self):
        self.elements = []
        self.entry_count = 0

    def add(self, item, priority):
        heapq.heappush(self.elements,
                    (priority, self.entry_count, item))
        self.entry_count += 1

    def pop(self):
        priority, count, item = heapq.heappop(self.elements)
        return item, priority
```

Node structure becomes more important in UCS:

```python
class Node:
    def __init__(self, state, parent=None, action=None,
                 path_cost=0):
        self.state = state
        self.parent = parent
        self.action = action
        self.path_cost = path_cost
```

Figure 6.7 provides an overview of the key advantages and disadvantages of UCS. It highlights UCS's ability to find optimal solutions while also pointing out its potentially high memory usage:

Advantages:

1. Solution Quality

   - Guarantees optimal solutions;
   - Effectively handles varying action costs;

2. Search Behavior

   - Explores promising paths first;
   - Naturally adapts to cost distributions;

3. Applicability

   - Ideal for problems with varying action costs;
   - Suitable for route-finding with different travel times;

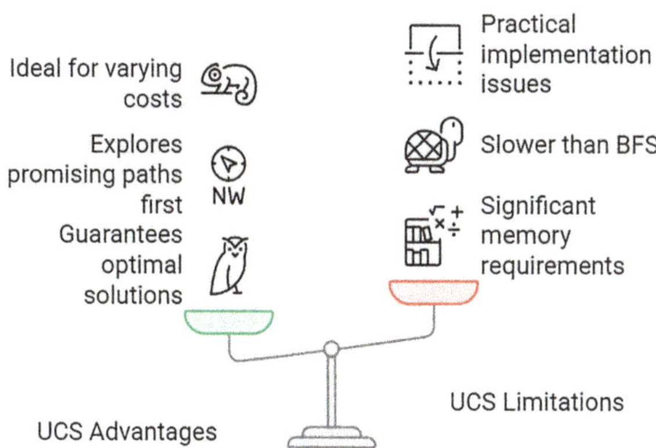

Fig. 6.7 Strengths and limitations of Uniform-Cost Search

Limitations:

1. Memory Requirements

   - Must store all nodes with cost less than solution;
   - Can be significant for problems with many similar-cost paths;

2. Performance Considerations

   - Slower than BFS for uniform-cost actions;
   - Priority queue operations add computational overhead;

3. Practical Issues

   - Requires careful implementation of the priority queue;
   - May explore many nodes with similar costs.

This summary helps us understand the trade-offs involved in using UCS for problems with varying action costs.

Implementation Considerations

Several practical aspects require attention:

1. Cost Updates:

```python
def update_cost(self, item, new_priority):
    # Remove old entry
    self.elements = [(p,c,i) for p,c,i in self.elements
                     if i != item]
    heapq.heapify(self.elements)
    # Add new entry
    self.add(item, new_priority)
```

2. State Comparison:

```python
def compare_states(state1, state2):
    # Implement proper state comparison
    # Consider hash values for efficiency
    return state1 == state2
```

UCS proves particularly valuable in:

- Navigation systems with varying road speeds;
- Network routing with different link costs;
- Resource allocation with varying costs.

This algorithm forms the basis for more advanced informed search methods like A* search, which adds heuristic information to further improve search efficiency. The next section will explore practical applications and common variations of these basic search algorithms.

6.4   Implementation Considerations

Effective implementation of search algorithms requires careful attention to data structures and implementation strategies. This section explores practical aspects of implementing search algorithms in Python, focusing on efficiency and reliability.

6.4.1   Basic Data Structures

The choice of data structures significantly impacts search algorithm performance. Each basic search algorithm requires its own specific data structure:

```python
# Queue implementation for BFS
from collections import deque
class Queue:
    def __init__(self):
        self.items = deque()

    def add(self, item):
        self.items.append(item)

    def pop(self):
        return self.items.popleft()

    def is_empty(self):
        return len(self.items) == 0

# Stack implementation for DFS
class Stack:
    def __init__(self):
        self.items = []

    def push(self, item):
        self.items.append(item)

    def pop(self):
        return self.items.pop()

    def is_empty(self):
        return len(self.items) == 0
```

6.4.2 *Priority Queue for UCS*

Uniform-Cost Search requires a more sophisticated priority queue implementation:

```python
import heapq


class PriorityQueue:
    def __init__(self):
        self.elements = []
        self.entry_count = 0

    def add(self, item, priority):
        heapq.heappush(self.elements,
                       (priority, self.entry_count, item))
        self.entry_count += 1

    def pop(self):
        priority, count, item = heapq.heappop(self.elements)
        return item, priority

    def update_priority(self, item, new_priority):
        for i, (p, c, itm) in enumerate(self.elements):
            if itm == item:
                self.elements[i] = (new_priority, c, itm)
                heapq.heapify(self.elements)
                return
```

6.4.3 *State Representation*

Proper state representation is crucial for efficient search:

```python
class State:
    def __init__(self, data):
        self.data = data

    def __eq__(self, other):
        return self.data == other.data

    def __hash__(self):
        return hash(str(self.data))
```

```python
def get_successors(self):
    # Generate successor states
    successors = []
    # Implementation depends on problem domain
    return successors
```

6.4.4   Memory Management Strategies

Efficient memory management is essential for handling large search spaces:

```python
class SearchNode:
    def __init__(self, state, parent=None, action=None,
                 path_cost=0):
        self.state = state
        self.parent = parent  # Store parent reference
        self.action = action  # Action that led to this state
        self.path_cost = path_cost

    def get_path(self):
        # Reconstruct path from root
        path = []
        current = self
        while current:
            path.append(current.state)
            current = current.parent
        return list(reversed(path))
```

6.4.5   Explored Set Implementation

The explored set helps prevent revisiting states:

```python
class ExploredSet:
    def __init__(self):
        self.explored = set()

    def add(self, state):
        self.explored.add(hash(state))

    def contains(self, state):
        return hash(state) in self.explored
```

6.4.6   Complete Search Implementation

Here's a complete implementation incorporating all components:

```python
def search(initial_state, goal_test, frontier):
    node = SearchNode(initial_state)
    frontier.add(node, 0)  # Priority only matters for UCS
    explored = ExploredSet()

    while not frontier.is_empty():
        current = frontier.pop()

        if goal_test(current.state):
            return current.get_path()

        explored.add(current.state)


        for successor in current.state.get_successors():
            if not explored.contains(successor):
                child = SearchNode(
                    successor,
                    parent=current,
                    path_cost=current.path_cost + 1
                )
                frontier.add(child)

    return None  # No solution found
```

6.4.7   Implementation Tips

Several practical considerations improve implementation quality:

1. Error Handling: Implement robust error checking for invalid states and actions.
2. Memory Efficiency: Use weak references for parent pointers when possible.
3. Performance Optimization: Choose appropriate data structures based on problem size.
4. Testing: Implement unit tests for basic operations and edge cases.

These implementation details form the foundation for building reliable and efficient search algorithms. The next section will explore practical applications and examples of these implementations in solving real-world problems.

Each implementation detail plays a crucial role in the overall performance and reliability of search algorithms. Careful attention to these aspects ensures efficient and correct operation when solving practical problems.

6.5 Practical Applications

Uninformed search algorithms find wide application across various domains. This section explores practical implementations through concrete examples, demonstrating how these algorithms solve real-world problems.

6.5.1 Puzzle Solving

The 8-puzzle provides an excellent example for understanding search algorithms in practice. Here's an implementation demonstrating uninformed search for puzzle solving:

```python
class PuzzleState:
    def __init__(self, board):
        self.board = board
        self.size = 3  # 3x3 grid

    def get_blank_position(self):
        for i in range(self.size):
            for j in range(self.size):
                if self.board[i][j] == 0:
                    return (i, j)

    def get_successors(self):
        successors = []
        i, j = self.get_blank_position()
        moves = [(0,1), (1,0), (0,-1), (-1,0)]  # Right, Down,
Left, Up

        for dx, dy in moves:
            new_i, new_j = i + dx, j + dy
            if 0 <= new_i < self.size and 0 <= new_j < self.size:
                new_board = [row[:] for row in self.board]
                new_board[i][j] = new_board[new_i][new_j]
                new_board[new_i][new_j] = 0
                successors.append(PuzzleState(new_board))

        return successors
```

6.5.2   *Route Finding*

Route finding represents a common application of search algorithms. Here's an implementation for grid-based navigation:

```python
class GridWorld:
    def __init__(self, grid):
        self.grid = grid
        self.rows = len(grid)
        self.cols = len(grid[0])

    def get_neighbors(self, position):
        x, y = position
        neighbors = []
        directions = [(0,1), (1,0), (0,-1), (-1,0)]

        for dx, dy in directions:
            new_x, new_y = x + dx, y + dy
            if (0 <= new_x < self.rows and
                0 <= new_y < self.cols and
                not self.grid[new_x][new_y]):
                neighbors.append((new_x, new_y))

        return neighbors


    def find_path(self, start, goal):
        return breadth_first_search(start, goal, self)
```

6.5.3   *Game Playing*

Search algorithms form the basis for game-playing strategies. Here's an example implementation for a simple game:

```python
class GameState:
    def __init__(self, board, player):
        self.board = board
        self.player = player

    def get_legal_moves(self):
        moves = []
        for i in range(3):
```

```
            for j in range(3):
                if self.board[i][j] == 0:
                    moves.append((i, j))
        return moves

    def make_move(self, move):
        i, j = move
        new_board = [row[:] for row in self.board]
        new_board[i][j] = self.player
        return GameState(new_board, -self.player)

    def is_terminal(self):
        # Check for win conditions
        return check_win(self.board) or len(self.get_legal_
moves()) == 0
```

6.5.4   Real-World Applications

Urban Navigation System:

```
class CityMap:
    def __init__(self, streets, traffic_data):
        self.streets = streets
        self.traffic_data = traffic_data

    def get_travel_time(self, start, end):
        return self.traffic_data[(start, end)]

    def find_fastest_route(self, start, goal):
        return uniform_cost_search(start, goal, self)
```

   Robot Path Planning:

```
class RobotEnvironment:
    def __init__(self, obstacles):
        self.obstacles = obstacles

    def is_valid_position(self, position):
        x, y = position
        return not any(self.collides(x, y, obs)
                    for obs in self.obstacles)

    def find_safe_path(self, start, goal):
        return breadth_first_search(start, goal, self)
```

6.5.5 *Performance Considerations*

Each application domain presents unique challenges:

1. State Space Size

    - Puzzle solving: $9!$ states for 8-puzzle
    - Route finding: Often millions of possible states
    - Game playing: Exponential growth with game depth

2. Memory Requirements Typical memory usage for different applications:

    - Small puzzles: A few megabytes
    - City navigation: Hundreds of megabytes
    - Complex games: Multiple gigabytes

3. Time Constraints Practical response time requirements:

    - Puzzle solving: Seconds acceptable
    - Route finding: Near real-time needed
    - Game playing: Often time-limited moves

These practical applications demonstrate how search algorithms solve real problems. Understanding these implementations helps in adapting the algorithms to new domains while maintaining efficiency and reliability.

The choice of algorithm depends heavily on the specific requirements of each application:

- BFS: Best for finding shortest paths in mazes
- DFS: Useful for exploring game trees
- UCS: Ideal for route finding with varying costs

The next section will provide exercises and case studies to reinforce these practical applications through hands-on experience.

6.6 Practice and Exercises

This section provides hands-on exercises to reinforce your understanding of the algorithms and concepts discussed in this chapter. Through practical implementation and experimentation, you will gain valuable experience in applying these techniques to real-world problems. Each exercise is designed to build your skills progressively, starting with basic implementations and moving toward more advanced applications.

The exercises are structured to help you:

- Understand the core principles of the algorithms.
- Implement and test the algorithms in Python.

- Analyze and visualize the results.
- Compare different approaches and their performance.

All exercises are available in an interactive Python notebook, which you can access using the QR code below or at: https://colab.research.google.com/drive/1RjkST4xRt2dTpWfung8ZWy_BGIjQWfl7.

The notebook includes example code, test cases, and visualization tools to help you complete the exercises effectively. Solutions to the exercises are available to instructors through the publisher's resources.

6.6.1   Practice and Programming Exercises

Exercise 1: Basic Grid Navigation with BFS

Learning Objectives:
- Understand the core components of Breadth-First Search (BFS).
- Implement BFS for grid navigation.
- Visualize the search process and results.

Tasks:
- Implement the BFS algorithm for a grid-based navigation problem.
- Use the provided grid with obstacles to test your implementation.
- Visualize the search process, showing explored nodes and the final path.
- Analyze the path length and the number of nodes explored.

Expected Outcomes:
- A working BFS implementation that finds a path from the start to the goal.
- A visualization of the search process and the final path.
- A brief analysis of the algorithm's performance.

Exercise 2: Depth-First Search (DFS) Implementation

Learning Objectives:
- Understand the differences between BFS and DFS.
- Implement DFS for grid navigation.
- Compare DFS and BFS in terms of path length and exploration.

Tasks:
- Implement the DFS algorithm for the same grid used in Exercise 1.
- Visualize the search process and the final path.
- Compare the results of DFS and BFS in terms of path length and the number of nodes explored.

Expected Outcomes:
- A working DFS implementation that finds a path from the start to the goal.
- A visualization of the search process and the final path.
- A comparison of DFS and BFS performance.

Exercise 3: Uniform-Cost Search (UCS) Implementation

Learning Objectives:
- Understand the principles of Uniform-Cost Search (UCS).
- Implement UCS for grid navigation with varying costs.
- Analyze the impact of cost on the search process.

Tasks:
- Implement UCS for a grid where each cell has a different movement cost.
- Visualize the search process and the final path.
- Compare the results of UCS with BFS and DFS in terms of path cost and exploration.

Expected Outcomes:
- A working UCS implementation that finds the lowest-cost path.
- A visualization of the search process and the final path.
- A comparison of UCS, BFS, and DFS performance.

Exercise 4: Algorithm Comparison and Analysis

Learning Objectives:
- Compare the performance of BFS, DFS, and UCS.
- Analyze the trade-offs between path length, cost, and exploration.

Tasks:
- Run BFS, DFS, and UCS on the same grid with obstacles.
- Record the path length, cost, and number of nodes explored for each algorithm.

- Create visualizations to compare the results.
- Write a brief analysis of the trade-offs between the algorithms.

Expected Outcomes:
- A comparison table showing the performance of BFS, DFS, and UCS.
- Visualizations of the search processes and final paths for each algorithm.
- A written analysis of the trade-offs between the algorithms.

Exercise 5: Advanced Visualization and Parameter Testing

Learning Objectives:
- Explore the impact of grid size and obstacle density on algorithm performance.
- Develop advanced visualizations to analyze algorithm behavior.

Tasks:
- Test BFS, DFS, and UCS on grids of different sizes (e.g., 5×5, 10×10, 15×15).
- Vary the obstacle density (e.g., 10%, 30%, 50%) and observe the impact on performance.
- Create visualizations to show how grid size and obstacle density affect the algorithms.
- Write a brief report summarizing your findings.

Expected Outcomes:
- Visualizations showing the impact of grid size and obstacle density on algorithm performance.
- A written report summarizing the results and conclusions.

6.6.2   Self-Assessment Questions

This section provides a set of self-assessment questions to test your understanding of the algorithms and concepts covered in this chapter. These questions are designed to reinforce key concepts and help you evaluate your knowledge. Answer each question before checking the solutions provided.

Multiple Choice Questions

1. What is the main difference between BFS and DFS?

    (a) BFS uses a stack, while DFS uses a queue.
    (b) BFS explores nodes level by level, while DFS explores nodes deeply first.
    (c) BFS is faster than DFS.
    (d) DFS always finds the shortest path.

2. Which algorithm guarantees the shortest path in an unweighted grid?

   (a) DFS.
   (b) UCS.
   (c) BFS.
   (d) None of the above.

3. What is the primary advantage of UCS over BFS?

   (a) UCS uses less memory.
   (b) UCS finds the lowest-cost path in weighted grids.
   (c) UCS is faster than BFS.
   (d) UCS explores fewer nodes.

4. Which algorithm is best suited for memory-constrained environments?

   (a) BFS.
   (b) DFS.
   (c) UCS.
   (d) All of the above.

5. What is the role of a heuristic function in informed search algorithms?

   (a) It calculates the exact cost of the path.
   (b) It estimates the remaining cost to the goal.
   (c) It determines the memory usage.
   (d) It explores all possible paths.

6. Which algorithm is most efficient for finding the lowest-cost path in a
   weighted grid?

   (a) BFS.
   (b) DFS.
   (c) UCS.
   (d) None of the above.

7. What is the main disadvantage of DFS?

   (a) It uses too much memory.
   (b) It does not guarantee the shortest path.
   (c) It is slower than BFS.
   (d) It cannot handle obstacles.

8. Which algorithm explores nodes based on their cumulative cost?

   (a) BFS.
   (b) DFS.
   (c) UCS.
   (d) All of the above.

9. What is the key advantage of BFS over DFS?

    (e) BFS uses less memory.
    (f) BFS guarantees the shortest path in unweighted grids.
    (g) BFS is faster than DFS.
    (h) BFS can handle weighted grids.

10. Which algorithm is best for finding a path in a grid with varying movement costs?

    (a) BFS.
    (b) DFS.
    (c) UCS.
    (d) None of the above.

Answer Key

1. (b) BFS explores nodes level by level, while DFS explores nodes deeply first.
2. (c) BFS guarantees the shortest path in an unweighted grid.
3. (b) UCS finds the lowest-cost path in weighted grids.
4. (b) DFS is best suited for memory-constrained environments.
5. (b) A heuristic function estimates the remaining cost to the goal.
6. (c) UCS is most efficient for finding the lowest-cost path in a weighted grid.
7. (b) DFS does not guarantee the shortest path.
8. (c) UCS explores nodes based on their cumulative cost.
9. (b) BFS guarantees the shortest path in unweighted grids.
10. (c) UCS is best for finding a path in a grid with varying movement costs.

6.7   Conclusions and Future Directions

This chapter has explored uninformed search algorithms and their applications in problem solving. We examined three fundamental algorithms: Breadth-First Search (BFS), Depth-First Search (DFS), and Uniform-Cost Search (UCS). Each algorithm offers distinct advantages for specific problem scenarios.

BFS explores nodes level by level. It guarantees finding the shortest path in unweighted graphs. However, it often requires significant memory to store frontier nodes. DFS explores deeply along each branch before backtracking. It uses minimal memory but may not find optimal solutions. UCS extends BFS by considering path costs. It always finds the lowest-cost path but requires additional computational overhead.

The practical implementations demonstrated how these algorithms apply to real-world problems. Grid navigation examples showed how each algorithm traverses the search space differently. Performance comparisons revealed the trade-offs between memory usage, optimality, and execution time.

References

1. Nilsson, N.J.: 8—Uninformed Search. In: Nilsson, N.J. (ed.) Artificial Intelligence: A New Synthesis. pp. 129–138. Morgan Kaufmann, Oxford (1998). https://doi.org/10.1016/B978-0-08-049945-1.50014-9.
2. Ginsberg, M.: CHAPTER 4—HEURISTIC SEARCH. In: Ginsberg, M. (ed.) Essentials of Artificial Intelligence. pp. 68–85. Morgan Kaufmann, San Francisco (1993). https://doi.org/10.1016/B978-1-55860-221-2.50009-1.
3. Edelkamp, S.: Heuristic search: theory and applications. Waltham, MA: Morgan Kaufmann (2012).
4. Edelkamp, S., Schrödl, S.: Chapter 2—Basic Search Algorithms. In: Edelkamp, S. and Schrödl, S. (eds.) Heuristic Search. pp. 47–87. Morgan Kaufmann, San Francisco (2012). https://doi.org/10.1016/B978-0-12-372512-7.00002-X.
5. Nilsson, N.J.: 9—Heuristic Search. In: Nilsson, N.J. (ed.) Artificial Intelligence: A New Synthesis. pp. 139–162. Morgan Kaufmann, Oxford (1998). https://doi.org/10.1016/B978-0-08-049945-1.50015-0.
6. Newell, A., Shaw, J., Simon, H.: Report on a general problem-solving program. Presented at the IFIP Congress (1959).
7. Blocho, M.: Chapter 4—Heuristics, metaheuristics, and hyperheuristics for rich vehicle routing problems. In: Nalepa, J. (ed.) Smart Delivery Systems. pp. 101–156. Elsevier (2020). https://doi.org/10.1016/B978-0-12-815715-2.00009-9.
8. Artificial Intelligence: A Modern Approach, 4th US ed., https://aima.cs.berkeley.edu/, last accessed 2025/02/20.
9. BREADTH FIRST SEARCH—Artificial Intelligence, https://intelligence.worldofcomputing.net/ai-search/breadth-first-search.html, last accessed 2025/02/20.
10. Lim, K.L., Seng, K.P., Yeong, L.S., Ang, L.-M., Ch'ng, S.I.: Uninformed pathfinding: A new approach. Expert Systems with Applications. 42, 2722–2730 (2015). https://doi.org/10.1016/j.eswa.2014.10.046.
11. UNIFORM-COST SEARCH—Artificial Intelligence, https://intelligence.worldofcomputing.net/ai-search/uniform-cost-search.html, last accessed 2025/02/20.

Chapter 7
Informed Search Algorithms

Abstract This chapter presents informed search algorithms, which leverage domain-specific knowledge to guide problem-solving in artificial intelligence. These algorithms extend uninformed search methods by incorporating heuristic information. This approach dramatically improves search efficiency while maintaining solution quality guarantees under certain conditions.

7.1 Foundations of Informed Search

Informed search algorithms represent a fundamental advancement in artificial intelligence, distinguishing themselves from uninformed methods through their strategic use of domain-specific knowledge [1, 2]. This section explores the core concepts underlying informed search and establishes their importance in modern AI applications.

7.1.1 Basic Concepts and Motivation

Informed search algorithms, also known as heuristic search methods [3, 4], incorporate additional knowledge about the problem domain to guide the search process more efficiently. The fundamental idea is straightforward: by using information about the goal, we can make better decisions about which paths to explore first.

Consider a simple navigation problem where an autonomous agent needs to find a path from point A to point B in a city. While an uninformed search algorithm would systematically explore all possible routes, an informed search algorithm might use the straight-line distance to the destination to prioritize more promising paths.

The general informed search problem can be formally defined as [1, 2]:

$$P = \left(S, A, T, c, s_0, G, h \right),$$

O. Kuznetsov, *Intelligent Systems: From Theory to Applications*, Cognitive Technologies, https://doi.org/10.1007/978-3-032-00044-6_7

where:

- S is the set of possible states;
- A is the set of available actions;
- $T : S \times A \to S$ is the transition function;
- $c : S \times A \times S \to R^+$ is the cost function;
- $s_0 \in S$ is the initial state;
- $G \subseteq S$ is the set of goal states;
- $h : S \to R^+$ is the heuristic function estimating cost to goal.

7.1.2   Comparison with Uninformed Search Methods

Informed search methods differ from uninformed approaches in several key aspects [1, 2]:

1. Search Efficiency: Uninformed methods like breadth-first or depth-first search explore the state space systematically but blindly. In contrast, informed search uses heuristic guidance to focus exploration on promising areas first.
2. Knowledge Utilization: While uninformed methods treat all unexplored paths equally, informed search leverages domain knowledge through heuristic functions to make educated guesses about which paths might lead to the goal more quickly.
3. Computational Resources: Informed methods typically examine fewer nodes to find a solution, though they may require additional computation per node to evaluate the heuristic function.

7.1.3   Role of Heuristic Functions

The heuristic function $h(n)$ plays a central role in informed search. It provides an estimate of the cost from any state to the nearest goal state. A good heuristic function has several important properties [1, 2]:

- $h(n)$ estimates the cost from node n to the goal;
- $h(g) = 0$ for any goal state g;
- $h(n) \geq 0$ for all nodes n.

For example, in route-finding problems, the straight-line distance serves as a natural heuristic function because [1, 2]:

- It never overestimates the actual driving distance;
- It is relatively simple to compute;
- It provides meaningful guidance toward the goal.

7.1.4 Requirements for Effective Heuristics

For a heuristic function to be effective in practice, it must satisfy certain requirements [1, 2]:

1. Admissibility: A heuristic $h(n)$ is admissible if it never overestimates the actual cost to reach the goal:

$$\forall n : h(n) \le h^*(n),$$

where $h^*(n)$ is the actual optimal cost from n to the goal.

2. Consistency (or Monotonicity): For any node n and its successor n' reached via action a:

$$h(n) \le c(n,a,n') + h(n').$$

3. Computational Efficiency: The heuristic function should be relatively inexpensive to compute, as it will be evaluated many times during the search process.
4. Informative Value: The heuristic should provide meaningful differentiation between states. A trivial heuristic that always returns zero, while admissible, offers no guidance to the search process.

These requirements establish a framework for developing and evaluating heuristic functions. The better a heuristic function satisfies these requirements while remaining computationally efficient, the more effective the informed search process will be.

The foundation of informed search rests on these core concepts. Understanding them is crucial for implementing effective search algorithms and applying them to real-world problems. The following sections will build upon these fundamentals to explore specific informed search algorithms and their practical applications.

7.2 Best-First Search Algorithms

Best-first search algorithms form a family of informed search strategies that select the most promising nodes for expansion based on an evaluation function [5]. This section examines the general framework and specific implementations of these powerful search methods.

7.2.1 General Best-First Search Framework

The best-first search strategy follows a simple yet powerful principle: always expand the most promising node according to a specified evaluation function [5]. This approach can be formalized through the following algorithm structure:

```
def
best_first_search(initial_state, goal_test, evaluation_function):
    frontier = PriorityQueue()   # Ordered by evaluation_function
    frontier.add(Node(initial_state))
    explored = set()

    while not frontier.is_empty():
        node = frontier.pop()   # Get best node

        if goal_test(node.state):
            return solution(node)

        explored.add(node.state)


        for child in expand(node):
            if child.state not in explored and
                child not in frontier:
                frontier.add(child)
            elif child in frontier:
                if evaluation_function(child) <
                    evaluation_function(frontier[child]):
                    frontier.replace(child)

    return failure
```

The algorithm maintains two main data structures:

1. A priority queue (frontier) containing unexpanded nodes;
2. A set of explored states to avoid cycles.

The evaluation function $f(n)$ determines the order in which nodes are explored. Different evaluation functions lead to different search behaviors.

7.2.2   Greedy Best-First Search

Greedy best-first search represents the simplest form of best-first search [6]. It uses the heuristic function directly as its evaluation function:

$$f(n) = h(n),$$

where $h(n)$ estimates the cost from node n to the goal.

This strategy always expands the node that appears closest to the goal. While this can lead to fast solutions, it has several important characteristics:

- It may not find the optimal solution;
- It can get stuck in local minima;
- It often performs well in problems where the heuristic is accurate.

The time complexity depends on the accuracy of the heuristic but can be $O(b^m)$ in the worst case, where b is the branching factor and m is the maximum depth.

7.2.3 A* Algorithm: Principles and Properties

A* algorithm combines the benefits of uniform-cost search and greedy best-first search [2, 6]. Its evaluation function is:

$$f(n) = g(n) + h(n),$$

where:

- $g(n)$ is the actual cost from the start node to node n;
- $h(n)$ is the estimated cost from node n to the goal.

The intuition behind A* is straightforward [2, 6]: $f(n)$ estimates the total cost of the cheapest solution path going through node n. The algorithm maintains a balance between:

- Following already discovered low-cost paths ($g(n)$);
- Exploring paths that appear to lead toward the goal ($h(n)$).

7.2.4 Theoretical Guarantees and Optimality Conditions

A* provides important theoretical guarantees under specific conditions [2, 6]:

1. Completeness: A* is complete if the branching factor is finite and every action has a cost greater than some small positive constant ϵ.
2. Optimality: A* is optimal if the heuristic function $h(n)$ is admissible. For any node n:

$$h(n) \leq h^*(n),$$

where $h^*(n)$ is the actual cost of the optimal path from n to the goal.
3. Consistency Condition: For any node n and successor n' with step cost $c(n, a, n')$:

$$h(n) \leq c(n,a,n') + h(n'),$$

when the heuristic is consistent, A* never needs to reopen closed nodes, leading to more efficient search.

4. Time and Space Complexity:

   - Time: $O(b^d)$ where b is the branching factor and d is the depth of the optimal solution;
   - Space: $O(b^d)$ as all nodes must be stored in memory.

The effectiveness of A* depends critically on the quality of the heuristic function:

- With $h(n) = 0$, A* reduces to uniform-cost search;
- With $h(n) = h^{(n)}$, A follows only optimal paths;
- Better heuristics lead to fewer node expansions.

These theoretical properties make A* the algorithm of choice for many informed search applications where finding optimal solutions is important. However, its memory requirements can be substantial, leading to the development of various memory-efficient variants which we will explore in later sections.

7.3 Practical Implementation Considerations

This section addresses key practical aspects of implementing informed search algorithms effectively. We explore essential considerations for real-world applications, focusing on concrete implementation strategies and optimizations.

7.3.1 Heuristic Function Design

The design of effective heuristic functions requires balancing accuracy with computational efficiency. Let's examine practical approaches to heuristic design through a pathfinding example.

For a grid-based pathfinding problem, common heuristic functions include:

```python
def manhattan_distance(current, goal):
    """Manhattan distance heuristic."""
    return abs(current[0] - goal[0]) + abs(current[1] - goal[1])
def euclidean_distance(current, goal):
    """Euclidean distance heuristic."""
    return ((current[0] - goal[0])**2 +
            (current[1] - goal[1])**2)**0.5
```

Key principles for heuristic design:

- Domain-specific insights should inform the heuristic. In pathfinding, geometric distances provide natural heuristics.
- The heuristic should be computationally inexpensive. Simple calculations are preferred.

- The heuristic must be admissible for A* optimality. Consider physical constraints when designing the function.

7.3.2   Data Structures for Efficient Implementation

Efficient implementation relies heavily on appropriate data structure choices. Here's a basic implementation of priority queue-based structures for informed search:

```python
import heapq
class PriorityQueue:
    def __init__(self):
        self.elements = []
        self.entry_count = 0

    def push(self, item, priority):
        heapq.heappush(self.elements,
                    (priority, self.entry_count, item))
        self.entry_count += 1

    def pop(self):
        return heapq.heappop(self.elements)[2]

    def empty(self):
        return len(self.elements) == 0
```

This implementation ensures:

- O(log n) time complexity for insertions and removals;
- Stable ordering for items with equal priorities;
- Efficient memory usage.

7.3.3   Memory Management Strategies

Memory management is crucial for handling large search spaces. Here are practical strategies:

1. Node Structure:

```python
class Node:
    def __init__(self, state, parent=None, action=None,
                 path_cost=0):
        self.state = state
```

```python
        self.parent = parent
        self.action = action
        self.path_cost = path_cost
        self.depth = 0 if parent is None else parent.depth + 1

    def __lt__(self, other):
        # Required for priority queue comparison
        return self.path_cost < other.path_cost
```

2. Explored Set Implementation:

```python
def explored_set():
    return set()   # For hashable states
```

3. Memory-Efficient State Representation:

```python
def state_to_tuple(state):
    """Convert state to immutable representation."""
    return tuple(state)   # For list/array states
```

7.3.4   Implementation Example: A* Pathfinding

Here's a complete example implementing A* for grid-based pathfinding:

```python
def a_star_search(grid, start, goal):
    frontier = PriorityQueue()
    frontier.push(Node(start), 0)
    came_from = {start: None}
    cost_so_far = {start: 0}

    while not frontier.empty():
        current = frontier.pop()

        if current.state == goal:
            return reconstruct_path(came_from, start, goal)

        for next_state in get_neighbors(current.state, grid):
            new_cost = (cost_so_far[current.state] +
                        movement_cost(current.state,
                                      next_state))
```

```
        if (next_state not in cost_so_far or
            new_cost < cost_so_far[next_state]):
            cost_so_far[next_state] = new_cost
            priority = new_cost + manhattan_distance(
                next_state, goal)
            frontier.push(Node(next_state), priority)
            came_from[next_state] = current.state

    return None
def reconstruct_path(came_from, start, goal):
    current = goal
    path = []

    while current != start:
        path.append(current)
        current = came_from[current]

    path.append(start)
    path.reverse()
    return path
```

Key optimization considerations (Fig. 7.1):

1. State Space Representation:

    - Use compact state representations;
    - Implement efficient state comparison operations;
    - Consider using bit operations for small state spaces;

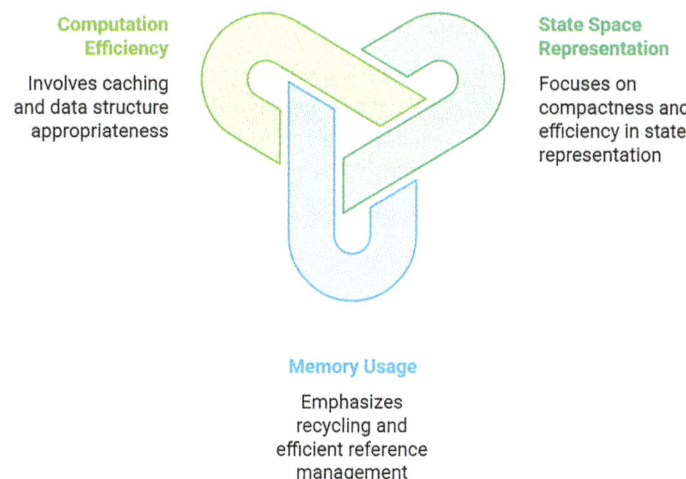

Fig. 7.1 Optimization strategies overview

2. Memory Usage:

   - Implement state recycling where possible;
   - Use generator functions for successor generation;
   - Consider using weak references for parent pointers;

3. Computation Efficiency:

   - Cache heuristic values when expensive to compute;
   - Implement incremental state updates;
   - Use appropriate data structures for the specific problem.

This implementation provides a practical foundation for informed search algorithms. The code can be adapted for different problem domains by modifying the state representation, heuristic function, and successor generation logic while maintaining the core algorithm structure.

These practical considerations significantly impact the performance of informed search implementations in real-world applications. The next section will explore advanced variants of these algorithms that address specific performance requirements and constraints.

7.4   Advanced Topics and Variants

This section explores advanced variants of informed search algorithms. These variants address specific challenges such as memory constraints, real-time requirements, and performance optimization in complex domains.

7.4.1   *IDA* (Iterative Deepening A*)*

IDA* combines the memory efficiency of iterative deepening with the heuristic guidance of A*. The algorithm performs a series of depth-first searches with an incrementally increasing cost bound.

The basic structure of IDA* is:

```
def ida_star(initial_state, goal_test, h):
    bound = h(initial_state)
    path = [initial_state]

    while True:
        t = search(path, 0, bound, goal_test, h)
        if t == FOUND: return path
        if t == INF: return FAILURE
        bound = t
```

The search function performs a depth-first search up to the current bound:

```python
def search(path, g, bound, goal_test, h):
    node = path[-1]
    f = g + h(node)
    if f > bound: return f
    if goal_test(node): return FOUND

    min_cost = INF
    for successor in expand(node):
        if successor not in path:
            path.append(successor)
            t = search(path, g + cost(node, successor),
                        bound, goal_test, h)
            if t == FOUND: return FOUND
            if t < min_cost: min_cost = t
            path.pop()

    return min_cost
```

Key advantages of IDA*:

- Memory usage is $O(d)$ where d is the solution depth;
- Maintains optimality guarantee of A* with admissible heuristics;
- Works well in domains with uniform step costs.

7.4.2 *Memory-Bounded Variants*

SMA* (Simplified Memory-Bounded A*) addresses memory limitations by maintaining a fixed-size set of nodes. When memory is full, it discards least promising nodes but retains information about their quality.

The core modification to A* is:

```python
def sma_star(initial_state, goal_test, h, memory_limit):
    frontier = LimitedPriorityQueue(memory_limit)
    frontier.push(Node(initial_state), h(initial_state))

    while not frontier.empty():
        if frontier.is_full():
            worst_node = frontier.remove_worst()
            # Store backup information about subtree
            backup_value = compute_backup(worst_node)
            store_backup(worst_node.parent, backup_value)
```

```
        node = frontier.pop()
        if goal_test(node.state): return solution(node)

        for child in expand(node):
            value = max(node.value, g(child) + h(child))
            frontier.push(child, value)


    return failure
```

7.4.3 *Bidirectional Search with Heuristics*

Bidirectional search simultaneously explores forward from the initial state and
backward from the goal state. With heuristics, this approach can be particularly
effective.

Implementation framework:

```
def bidirectional_a_star(start, goal, h_forward, h_backward):
    forward_frontier = PriorityQueue()
    backward_frontier = PriorityQueue()

    forward_frontier.push(start, h_forward(start))
    backward_frontier.push(goal, h_backward(goal))

    forward_reached = {start: 0}
    backward_reached = {goal: 0}

    while not (forward_frontier.empty() and
               backward_frontier.empty()):
        if meet_in_middle(forward_reached,
                          backward_reached):
            return construct_solution(forward_reached,
                                      backward_reached)

        if forward_frontier.f_min() <
           backward_frontier.f_min():
            expand_forward()
        else:
            expand_backward()


    return failure
```

7.4.4   *Real-Time Search Algorithms*

Real-time search algorithms must make decisions with limited computational resources. LRTA* (Learning Real-Time A*) is a popular example that updates heuristic estimates during search.

Basic LRTA* implementation:

```python
def lrta_star_step(current, goal, h):
    if current == goal: return None

    successors = expand(current)
    if not successors: return None

    # Update heuristic value
    h[current] = min(cost(current, s) + h[s]
                     for s in successors)

    # Choose best successor
    next_state = min(successors,
                     key=lambda s: cost(current, s) + h[s])

    return next_state
```

Key features of real-time search:

- Constant-time decision making;
- Learning from experience through heuristic updates;
- Interleaving planning and execution.

The choice among these variants depends on specific problem requirements (Fig. 7.2):

- Use IDA* when memory is severely constrained;
- Choose SMA* for problems with moderate memory constraints;
- Apply bidirectional search when good heuristics exist for both directions;
- Use real-time variants for dynamic environments requiring quick responses.

These advanced algorithms demonstrate how the basic principles of informed search can be adapted to meet various practical constraints while maintaining essential properties like completeness and optimality where possible.

7.5   Applications and Case Studies

This section examines practical applications of informed search algorithms across various domains. Through concrete examples, we explore how these algorithms solve real-world problems effectively.

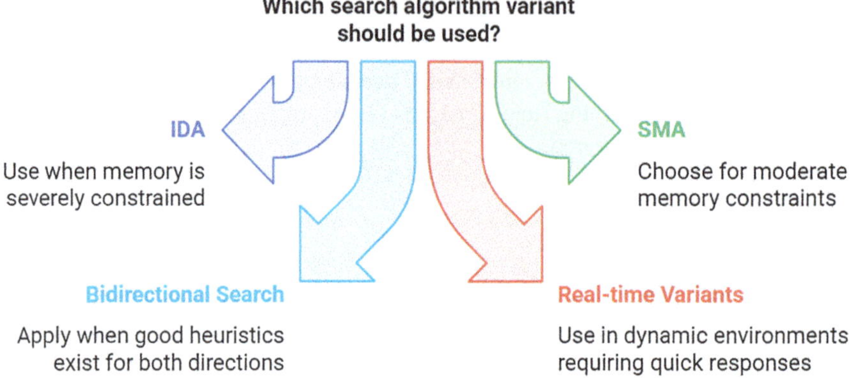

Fig. 7.2 Selecting a search algorithm

7.5.1 Path Planning and Navigation

Path planning represents one of the most common applications of informed search algorithms. Modern navigation systems rely heavily on variants of A* to find optimal routes.

Consider a city navigation problem:

```python
def city_navigation(map_graph, start, goal):
    def heuristic(node):
        return euclidean_distance(node, goal)

    def cost(current, next):
        return (road_distance(current, next) *
                traffic_factor(current, next))

    return a_star_search(map_graph, start, goal,
                         heuristic, cost)
```

Key considerations in navigation applications:

- Dynamic cost functions incorporating traffic conditions;
- Multiple heuristics for different optimization criteria;
- Real-time updates and replanning capabilities.

7.5.2 Game Playing

Informed search algorithms play a crucial role in game AI. Consider the 8-puzzle game implementation:

```python
class EightPuzzle:
    def __init__(self, board):
        self.board = board
        self.goal = [[1, 2, 3],
                     [4, 5, 6],
                     [7, 8, 0]]

    def heuristic(self):
        """Manhattan distance heuristic."""
        distance = 0
        for i in range(3):
            for j in range(3):
                if self.board[i][j] != 0:
                    x, y = divmod(self.board[i][j]-1, 3)
                    distance += abs(x-i) + abs(y-j)
        return distance

    def get_successors(self):
        # Generate valid moves
        x, y = self.find_empty()
        moves = []
        for dx, dy in [(0,1), (1,0), (0,-1), (-1,0)]:
            new_x, new_y = x + dx, y + dy
            if 0 <= new_x < 3 and 0 <= new_y < 3:
                new_board = [row[:] for row in self.board]
                new_board[x][y] = new_board[new_x][new_y]
                new_board[new_x][new_y] = 0
                moves.append(EightPuzzle(new_board))
        return moves
```

7.5.3   Puzzle Solving

Puzzle solving demonstrates the power of informed search algorithms in combinatorial problems. Let's examine the implementation for the 15-puzzle:

```python
def solve_15_puzzle(initial_state):
    def manhattan_distance(state):
        distance = 0
        size = 4  # 4x4 grid
        for i in range(size):
            for j in range(size):
                if state[i][j] != 0:
                    target_x = (state[i][j] - 1) // size
```

```
                      target_y = (state[i][j] - 1) % size
                      distance += (abs(target_x - i) +
                                   abs(target_y - j))
        return distance


    def ida_star_solve():
        bound = manhattan_distance(initial_state)
        path = [initial_state]

        while True:
            t = search(path, 0, bound)
            if t == FOUND: return path
            if t == INF: return None
            bound = t
```

7.5.4   Real-World Applications

Informed search algorithms find applications in various practical domains:

1. Robotics Path Planning:

```
    def robot_path_planning(environment, start, goal):
        def collision_free(point):
            return not environment.check_collision(point)

        def heuristic(state):
            if not collision_free(state):
                return float('inf')
            return euclidean_distance(state, goal)

        return a_star_search(start, goal, heuristic)
```

2. Network Routing:

```
    def network_routing(network, source, destination):
        def heuristic(node):
            return (network.bandwidth_estimate(node,
                                        destination) +
                    network.latency_estimate(node,
                                        destination))

        return a_star_search(network, source, destination,
                        heuristic)
```

3. Supply Chain Optimization:

```python
def optimize_delivery_route(warehouses, orders):
    def cost_heuristic(state):
        undelivered = get_undelivered_orders(state)
        return min_possible_delivery_cost(undelivered)

    return a_star_search(initial_state, goal_state,
                         cost_heuristic)
```

Each application domain presents unique challenges (Fig. 7.3).

These case studies demonstrate how informed search algorithms can be adapted to solve diverse real-world problems effectively. The key to successful application lies in:

- Proper problem formulation;
- Effective heuristic design;
- Careful implementation considering domain constraints;
- Appropriate algorithm selection based on requirements.

The versatility of informed search algorithms makes them invaluable tools in modern artificial intelligence applications.

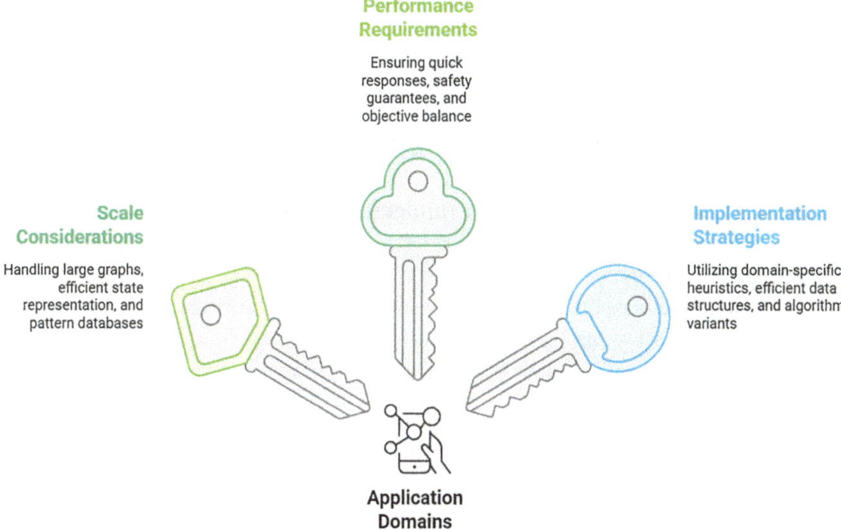

Fig. 7.3 Navigating challenges in diverse application domains

7.6 Practice and Exercises

This section provides hands-on exercises to reinforce understanding of informed search algorithms through practical implementation and experimentation. All exercises are available in an interactive Python notebook that can be accessed using the QR code below or at: https://colab.research.google.com/drive/1FerZnsernLEJUm9Y_k4g2CsYlNSb1de4

The notebook contains interactive Python code that you can run and modify to better understand genetic algorithms. Each exercise builds upon the previous ones, gradually introducing more complex aspects of genetic algorithm implementation. Solutions to the exercises are available to instructors through the publisher's resources.

7.6.1 Practice and Programming Exercises

Exercise 1: Basic Best-First Search Implementation

Learning Objectives:
- Understand the core components of best-first search algorithms;
- Implement a basic search algorithm with visualization;
- Learn to handle search states and transitions effectively.

Tasks:
- Implement the Node class for search state representation;
- Develop the best-first search algorithm with visualization;
- Test the implementation on a simple grid navigation problem;
- Analyze the search process using provided visualization tools.

Exercise 2: A* Algorithm Implementation

Learning Objectives
- Master the A* algorithm implementation;
- Understand the role of heuristic functions;
- Learn to visualize and analyze search performance.

Tasks
- Implement A* search with multiple heuristic functions;
- Create visualization for the search process;
- Compare different heuristic functions' performance;
- Analyze the impact of different grid configurations.

Exercise 3: Advanced Algorithm Variants

Learning Objectives:
- Understand memory-efficient search variants;
- Implement IDA* and SMA* algorithms;
- Compare algorithm performance characteristics.

Tasks:
- Implement IDA* algorithm with iterative deepening;
- Develop SMA* with memory constraints;
- Create performance comparison framework;
- Analyze trade-offs between different algorithms.

Exercise 4: Performance Analysis

Learning Objectives:
- Learn to measure algorithm performance metrics;
- Understand scaling behavior of search algorithms;
- Develop skills in comparative analysis.

Tasks:
- Implement performance measurement framework;
- Test algorithms on different grid sizes;
- Create visualization for performance metrics;
- Analyze and interpret performance results.

Exercise 5: Multi-Agent Pathfinding

Learning Objectives:
- Understand multi-agent pathfinding challenges;
- Learn to coordinate multiple search instances;
- Develop visualization for multi-agent scenarios.

Tasks:
- Implement multi-agent pathfinding environment;
- Develop path coordination mechanisms;
- Create interactive visualization system;
- Analyze agent interaction patterns.

Each Exercise Includes:
- A step-by-step implementation guide;
- Example code and test cases;
- Visualization tools for analysis;
- Discussion questions for deeper understanding.

Key Implementation Considerations:
- Proper handling of search states and transitions;
- Efficient data structures for frontier management;
- Effective visualization of search progress;
- Accurate performance measurement;
- Clear code organization and documentation.

The exercises progressively build upon each other, leading to a comprehensive understanding of informed search algorithms and their practical applications.

Note: Students should refer to the lecture materials and textbook sections covered in this chapter while working through these exercises. The accompanying Python notebook provides additional guidance and helper functions.

7.6.2 Self-Assessment Questions

Before attempting these questions, review the chapter material thoroughly, complete all practical exercises, and understand both theoretical concepts and their implementations. Take time to answer each question before checking the solutions.

Multiple Choice Questions

1. What is the key difference between informed and uninformed search algorithms?

    (a) Informed search uses more memory
    (b) Informed search uses domain-specific knowledge to guide the search
    (c) Informed search is always faster
    (d) Informed search guarantees optimal solutions

2. Which statement about A* algorithm is correct?

    (a) It always finds the shortest path regardless of the heuristic
    (b) It is guaranteed to find the optimal path if the heuristic is admissible

   (c) It requires less memory than breadth-first search
   (d) It never explores more nodes than depth-first search

3. What is an admissible heuristic?

   (a) A heuristic that never overestimates the cost to the goal
   (b) A heuristic that always finds the optimal path
   (c) A heuristic that uses minimal memory
   (d) A heuristic that explores all possible paths

4. When comparing IDA* to A*, which statement is true?

   (a) IDA* always finds shorter paths than A*
   (b) IDA* uses less memory but may re-explore states
   (c) IDA* is always faster than A*
   (d) IDA* requires more memory than A*

5. What is the main advantage of using SMA* over regular A*?

   (a) It finds paths faster
   (b) It guarantees shorter paths
   (c) It bounds memory usage
   (d) It explores fewer nodes

6. In the context of multi-agent pathfinding, what is a key challenge?

   (a) Finding optimal paths for each agent independently
   (b) Coordinating paths to avoid conflicts between agents
   (c) Using less memory than single-agent pathfinding
   (d) Implementing efficient heuristics

7. When is Greedy Best-First Search preferable to A*?

   (a) When memory is limited
   (b) When optimal solutions are required
   (c) When quick, potentially suboptimal solutions are acceptable
   (d) When the search space is very small

8. What role does the heuristic function play in informed search?

   (a) It determines the exact cost of the path
   (b) It estimates the remaining cost to the goal
   (c) It calculates the total path length
   (d) It determines the memory usage

9. In pathfinding applications, why is Manhattan distance often used as a heuristic?

   (a) It is always optimal
   (b) It is easy to compute and never overestimates
   (c) It considers diagonal movements
   (d) It accounts for obstacles

10. What makes a search algorithm complete?

    (a) It always finds the optimal solution
    (b) It uses minimal memory
    (c) It guarantees finding a solution if one exists
    (d) It explores all possible paths

Answer Key

1. (b) Informed search uses domain-specific knowledge to guide the search process more effectively.
2. (b) A* is guaranteed to find the optimal path if the heuristic is admissible.
3. (a) An admissible heuristic never overestimates the cost to reach the goal.
4. (b) IDA* uses less memory but may re-explore states multiple times.
5. (c) SMA* bounds memory usage while still providing optimal solutions when possible.
6. (b) Coordinating paths to avoid conflicts between agents is a key challenge in multi-agent pathfinding.
7. (c) When quick, potentially suboptimal solutions are acceptable, Greedy Best-First Search can be preferable.
8. (b) The heuristic function estimates the remaining cost to the goal, guiding the search process.
9. (b) Manhattan distance is easy to compute and never overestimates the actual cost, making it an effective heuristic.
10. (c) A search algorithm is complete if it guarantees finding a solution whenever one exists.

7.7   Conclusions and Future Directions

Informed search algorithms provide effective solutions to many complex problems. This chapter has shown how domain-specific knowledge dramatically improves search efficiency. We have examined both theoretical foundations and practical implementations. These algorithms demonstrate excellent performance across various applications.

Several key insights emerge from this chapter. First, informed search algorithms efficiently explore large solution spaces using heuristic guidance. Second, heuristic functions play a crucial role in directing the search toward optimal solutions. Third, variants like A*, IDA*, and SMA* offer different tradeoffs between memory usage and computational efficiency. Fourth, real-world applications demonstrate the versatility of these algorithms.

References

1. Blocho, M.: Chapter 4—Heuristics, metaheuristics, and hyperheuristics for rich vehicle routing problems. In: Nalepa, J. (ed.) Smart Delivery Systems. pp. 101–156. Elsevier (2020). https://doi.org/10.1016/B978-0-12-815715-2.00009-9.
2. Edelkamp, S.: Heuristic search: theory and applications. Waltham, MA: Morgan Kaufmann (2012).
3. Nilsson, N.J.: 9—Heuristic Search. In: Nilsson, N.J. (ed.) Artificial Intelligence: A New Synthesis. pp. 139–162. Morgan Kaufmann, Oxford (1998). https://doi.org/10.1016/B978-0-08-049945-1.50015-0.
4. Ginsberg, M.: CHAPTER 4—HEURISTIC SEARCH. In: Ginsberg, M. (ed.) Essentials of Artificial Intelligence. pp. 68–85. Morgan Kaufmann, San Francisco (1993). https://doi.org/10.1016/B978-1-55860-221-2.50009-1.
5. Norvig, P.: Artificial intelligence: a modern approach. Global edition. Pearson, Boston (2021).
6. Pearl, J.: Heuristics: intelligent search strategies for computer problem solving. Addison-Wesley Longman Publishing Co., Inc., USA (1984).

Chapter 8
The A* Algorithm

Abstract This chapter examines the A* algorithm in both theory and practice. It presents core principles and mathematical foundations of this essential pathfinding technique. Implementation strategies and algorithm variants receive detailed attention. We explore practical applications in robotics, gaming, and navigation systems. Memory optimization techniques and performance analysis tools help students optimize implementations. Hands-on Python examples demonstrate algorithm behavior in various scenarios. Students learn to adapt A* for different environments, from simple grids to complex terrain. The chapter covers advanced topics including parallel implementations and integration with modern AI.

8.1 Foundations of A*

The A* algorithm represents a fundamental milestone in artificial intelligence and computer science [1]. It combines the completeness of Dijkstra's algorithm [2, 3] with the efficiency of heuristic search methods. Let's explore its key principles and mathematical foundations.

8.1.1 Basic Principles and Mathematical Formulation

A* is built on a simple yet powerful idea. For each node n in the search space, it maintains an evaluation function [4]:

$$f(n) = g(n) + h(n),$$

where:

- $g(n)$ represents the actual cost from the start node to node n;
- $h(n)$ represents the estimated cost from node n to the goal.

O. Kuznetsov, *Intelligent Systems: From Theory to Applications*, Cognitive
Technologies, https://doi.org/10.1007/978-3-032-00044-6_8

The algorithm maintains two sets [4]:

1. OPEN: contains nodes that have been discovered but not yet expanded;
2. CLOSED: contains nodes that have been expanded.

At each iteration, A* selects the node with the lowest $f(n)$ value from the OPEN set. This selection process ensures that the algorithm explores the most promising paths first.

8.1.2   Role in Informed Search Strategies

A* belongs to the family of informed search algorithms [5]. Unlike uninformed methods like breadth-first or depth-first search, A* uses problem-specific knowledge encoded in its heuristic function [5, 6]. This knowledge guides the search toward the goal more efficiently.

The power of A* lies in its ability to combine [1, 5]:

- Actual path costs (like Dijkstra's algorithm);
- Heuristic estimates (like greedy best-first search).

This combination creates a balanced approach that is both efficient and optimal under certain conditions.

8.1.3   Comparison with Other Search Algorithms

Let's examine how A* compares to other common search algorithms [4].

Dijkstra's Algorithm

- Uses only $g(n)$ for path evaluation;
- Guarantees shortest path;
- Explores nodes in all directions.

Greedy Best-First Search

- Uses only $h(n)$ for evaluation;
- Often faster but not optimal;
- May get stuck in local minima.

A* effectively combines the advantages of both approaches while minimizing their drawbacks.

8.1.4 Properties of Admissibility and Optimality

A* possesses two crucial properties that make it particularly valuable [5].

Admissibility [5]

A heuristic $h(n)$ is admissible if it never overestimates the actual cost to reach the goal:

$$\forall n : h(n) \le h^*(n),$$

where $h^*(n)$ is the actual minimal cost from n to the goal.

Consistency (or Monotonicity) [5]

A heuristic is consistent if for every node n and successor n':

$$h(n) \le c(n,n') + h(n'),$$

where $c(n,n')$ is the cost of reaching n' from n.

Theorem [5]

- If $h(n)$ is admissible, A* will find an optimal path to the goal.

These properties ensure that:

1. A* never expands nodes that cannot be part of an optimal solution;
2. The first path found to the goal is guaranteed to be optimal;
3. A* expands the minimum number of nodes necessary among all optimal algorithms using the same heuristic.

The algorithm achieves this efficiency by maintaining a best-first search strategy while using admissible heuristics to guide the search toward promising directions. This combination makes A* both complete and optimal when using an admissible heuristic.

Understanding these foundational concepts is crucial for implementing A* effectively and applying it to real-world problems. The next sections will build upon these principles to explore practical implementations and advanced variations of the algorithm.

8.2   Algorithm Components and Implementation

8.2.1   Core Algorithm Structure

The A* algorithm follows a systematic approach to find optimal paths. Let's examine its core components and structure through a clear implementation.

The basic A* algorithm can be expressed in pseudocode:

```
def a_star(start, goal, h):
  open_set = {start}     # Nodes to be evaluated
  closed_set = {}        # Already evaluated nodes
  g_score = {start: 0}   # Cost from start to current node
  f_score = {start: h(start)}   # Estimated total cost
  while open_set is not empty:
      current = node in open_set with lowest f_score
      if current == goal:
          return reconstruct_path(current)

      open_set.remove(current)
      closed_set.add(current)

      for neighbor in get_neighbors(current):
          if neighbor in closed_set:
              continue

          tentative_g = g_score[current] + cost(current, neighbor)


          if neighbor not in open_set:
              open_set.add(neighbor)
          elif tentative_g >= g_score[neighbor]:
              continue

          g_score[neighbor] = tentative_g
          f_score[neighbor] = g_score[neighbor] + h(neighbor)
```

This implementation demonstrates the core logic of A*. Each component serves a specific purpose in ensuring optimal pathfinding.

8.2.2   Heuristic Functions and Their Properties

A heuristic function $h(n)$ estimates the cost from node n to the goal. Let's examine common heuristic functions:

Manhattan Distance

- For grid-based movement:

$$h(n) = |\, x_1 - x_2 \,| + |\, y_1 - y_2 \,|.$$

Euclidean Distance

- For unrestricted movement:

$$h(n) = \sqrt{(x_1 - x_2)^2 + (y_1 - y_2)^2}.$$

Diagonal Distance

- For 8-directional movement:

$$h(n) = \max\left(\|, \|x_1 - x_2\|, \|\|\|, \|\|\|, \|y_1 - y_2\|, \|\right).$$

A good heuristic function should be:

- Fast to compute;
- Easy to understand;
- Admissible (never overestimating);
- Consistent (satisfying the triangle inequality).

8.2.3   *Open and Closed Sets Management*

Efficient management of open and closed sets is crucial for A*'s performance.
Here's an optimal implementation approach:

```python
class Node:
  def __init__(self, state, g_score=float('inf'), h_score=0):
      self.state = state
      self.g_score = g_score
      self.h_score = h_score
      self.f_score = g_score + h_score
      self.parent = None

  def __lt__(self, other):
      return self.f_score < other.f_score
```

The open set is typically implemented as a priority queue, using f_score as the priority:

```
from heapq import heappush, heappop
open_set = []  # Priority queue
heappush(open_set, start_node)
```

8.2.4   *Implementation Strategies and Data Structures*

For optimal A* implementation, consider these key data structures:

Priority Queue (Open Set)

```
class PriorityQueue:
  def __init__(self):
      self.elements = []

  def push(self, item, priority):
      heappush(self.elements, (priority, item))

  def pop(self):
      return heappop(self.elements)[1]
```

Hash Table (Closed Set)

```
closed_set = set()  # For O(1) lookup time
```

Path Reconstruction

```
def reconstruct_path(current):
  path = []
  while current.parent:
      path.append(current)
      current = current.parent
  path.append(current)
  return path[::-1]
```

Memory Optimization

To reduce memory usage, consider:

- Using primitive data types where possible;
- Implementing a node pool for frequently created/destroyed nodes;
- Pruning the search space using domain-specific knowledge.

Time Complexity

- Worst case: $O(b^d)$, where b is the branching factor and d is the depth;
- Average case with good heuristic: $O(d)$.

Space Complexity

- Worst case: $O(b^d)$;
- Can be reduced using iterative deepening variants.

These implementation details ensure efficient A* operation while maintaining its optimality guarantees. The next section will explore various optimizations and adaptations for specific problem domains.

8.3   A* Variants and Extensions

While standard A* is powerful, practical applications often require modifications to handle specific constraints. This section explores major variants that address different limitations of the basic algorithm.

8.3.1   IDA* (Iterative Deepening A*)

IDA* combines the ideas of iterative deepening with A*'s heuristic approach. It addresses the memory limitations of A* while maintaining optimality guarantees.
   Basic Structure:

```
def ida_star(start, goal, h):
  threshold = h(start)
  while True:
      result = search(start, 0, threshold, goal, h)
      if result == FOUND: return threshold
      if result == infinity: return NOT_FOUND
      threshold = result
```

The main iteration performs depth-first search bounded by a threshold. At each iteration:

- Start with threshold = h(start);
- Perform depth-first search, cutting off when f(n) > threshold;
- If goal not found, increase threshold to minimum f-value that exceeded current threshold.

Time Complexity: $O(b^d)$.

Space Complexity: $O(d)$, where d is solution depth.

8.3.2   Memory-Bounded Variants

SMA* (Simplified Memory-Bounded A*) maintains optimal behavior while working within fixed memory constraints.

Key Features:

- Operates with a fixed-size node list;
- When memory is full, drops least promising nodes;
- Maintains information about dropped nodes;
- Can recover optimal paths despite memory limitations.

Implementation Approach:

```python
def sma_star(start, goal, h, max_nodes):
  open_set = PriorityQueue(max_size=max_nodes)
  open_set.push(start, h(start))
  while not open_set.empty():
      if open_set.full():
          remove_worst_leaf(open_set)
      current = open_set.pop()
      # Standard A* expansion with memory checks
```


8.3.3   Real-Time Variants

Real-Time A* (RTA*) focuses on quick decision-making rather than complete path planning.

Main Characteristics:

- Makes decisions based on local information;
- Updates heuristic values during search;
- Guarantees a move within fixed time;
- May not find optimal paths but acts quickly.

Basic Algorithm:

```
def rta_star(start, goal, h, look_ahead):
  current = start
  while current != goal:
      neighbors = get_neighbors(current)
      next_node = select_move(neighbors, look_ahead)
      update_heuristic(current, neighbors)
      current = next_node
```

8.3.4   Anytime Implementations

ARA* (Anytime Repairing A*) provides a framework for finding solutions with varying quality-time tradeoffs.
   Key Concepts:

- Starts with an inflated heuristic: $h'(n) = \epsilon \cdot h(n)$;
- Quickly finds sub-optimal solutions;
- Gradually reduces ϵ to improve solution quality;
- Reuses previous search efforts.

Implementation Structure:

```
def ara_star(start, goal, h, initial_epsilon):
  epsilon = initial_epsilon
  while epsilon > 1.0 and time_remains():
      solution = weighted_a_star(start, goal, h, epsilon)
      if solution:
          improve_solution(solution)
      epsilon = reduce_epsilon(epsilon)
```

Performance Characteristics:

- First solution found quickly with high ϵ;
- Solution quality improves as ϵ decreases;
- Final solution approaches optimality as $\epsilon \to 1$;
- Efficient reuse of previous search efforts.

Time and Quality Tradeoffs:

$$f(n) = g(n) + \epsilon \cdot h(n)$$

- Larger ϵ: Faster solutions, less optimal;
- Smaller ϵ: Slower solutions, more optimal;
- $\epsilon = 1$: Standard A* behavior.

Each variant addresses specific limitations of standard A*:

- IDA*: Memory constraints;
- SMA*: Fixed memory bounds;
- RTA*: Real-time requirements;
- ARA*: Anytime behavior.

These modifications maintain the core principles of A* while adapting to practical constraints. Understanding these variants is crucial for selecting the appropriate algorithm for specific application requirements.

8.4   Practical Applications

A* and its variants find extensive use in real-world applications [7–11]. This section explores concrete implementations across different domains.

8.4.1   Robotics and Autonomous Navigation

In robotics, A* serves as a fundamental tool for path planning. Let's examine a typical implementation for a mobile robot.

Basic Robot Navigation

```
def robot_navigation(start_pos, goal_pos, obstacles):
  def heuristic(pos):
      return euclidean_distance(pos, goal_pos)

  def get_valid_moves(pos):
      moves = []
      for dx, dy in [(0,1), (1,0), (0,-1), (-1,0)]:
          new_pos = (pos[0] + dx, pos[1] + dy)
          if is_valid(new_pos) and not_in_obstacle(new_pos):
              moves.append(new_pos)
      return moves
```

Dynamic Environment Handling

```
def dynamic_path_planning(robot_pos, goal, sensor_data):
  path = ara_star(robot_pos, goal, heuristic)
```

```
while not reached_goal(robot_pos, goal):
    if path_blocked(path, sensor_data):
        path = replan_path(robot_pos, goal, sensor_data)
    move_robot(next_step(path))
```

8.4.2   Game Development and AI

A* plays a crucial role in video game AI for pathfinding and strategic decision-making.

Game Map Navigation

```
class GameMap:
  def __init__(self, terrain_data):
      self.terrain = terrain_data
      self.movement_costs = {
          'grass': 1.0,
          'water': 2.5,
          'mountain': 3.0
      }
  def get_cost(self, pos1, pos2):
      terrain_type = self.terrain[pos2]
      return self.movement_costs[terrain_type]
```

NPC Movement Example

```
def npc_pathfinding(npc_pos, target_pos, game_map):
  def h(pos):
      # Consider terrain-specific costs
      base_distance = manhattan_distance(pos, target_pos)
      terrain_factor = game_map.get_terrain_factor(pos)
      return base_distance * terrain_factor
```

8.4.3   Route Planning Systems

Modern navigation systems use specialized versions of A* for efficient route calculation.

Highway Navigation System

```
class RoutePlanner:
  def __init__(self, road_network):
      self.network = road_network
      self.traffic_data = {}

  def calculate_route(self, start, destination, preferences):
      def h(location):
          return (
              self.distance_estimate(location, destination) *
              self.traffic_factor(location)
          )
```

Traffic Consideration

```
def adaptive_routing(start, end, traffic_data):
  cost_function = lambda x, y: (
      distance(x, y) * traffic_factor(x, y, traffic_data)
  )
  return a_star(start, end, cost_function)
```

8.4.4   Real-World Case Studies

Let's examine specific implementations in production systems.

Case 1: Warehouse Robot Navigation

```
class WarehouseRobot:
  def navigate_to_shelf(self, current_pos, target_shelf):
      path = a_star(
          start=current_pos,
          goal=target_shelf,
          h=manhattan_distance,
          constraints=self.warehouse_constraints
      )

      for movement in path:
          self.execute_movement(movement)
          self.check_obstacles()
```

Key Considerations:

- Real-time obstacle avoidance;
- Battery efficiency optimization;
- Multi-robot coordination.

Case 2: Urban Navigation System

```python
def city_navigation(start, end, transport_mode):
    if transport_mode == 'car':
        return vehicle_route(start, end)
    elif transport_mode == 'walk':
        return pedestrian_route(start, end)
    else:
        return multi_modal_route(start, end)
```

Implementation Challenges:

- Dynamic traffic conditions;
- Multiple transportation modes;
- Real-time updates;
- User preferences.

These practical applications demonstrate how A* adapts to different requirements:

- Memory efficiency in game environments;
- Real-time performance in robotics;
- Accuracy in navigation systems;
- Flexibility in mixed-mode transportation.

Each domain presents unique challenges that require specific modifications to the basic A* algorithm. Understanding these adaptations is crucial for implementing effective solutions in real-world scenarios.

8.5 Advanced Topics and Optimizations

This section explores advanced techniques for enhancing A* performance and adaptability in modern applications.

8.5.1 Performance Analysis and Optimization

Performance optimization of A* focuses on three key metrics: execution time, memory usage, and solution quality.

Time Complexity Analysis

The basic A* time complexity is: $O(b^d)$, where:

- b is the branching factor;
- d is the solution depth.

Memory Usage

Standard A* memory requirement: $O(b^d)$ for storing nodes.

Optimization Techniques

```python
class OptimizedNode:
    __slots__ = ['state', 'g_score', 'h_score', 'parent']
    def __init__(self, state, g_score=float('inf')):
        self.state = state
        self.g_score = g_score
        self.h_score = 0
        self.parent = None
```

Memory-Efficient Implementation

```python
def efficient_a_star(start, goal, h):
    open_set = PriorityQueue()
    visited = set()
    node_pool = NodePool(max_size=10000)   # Reuse node objects
    current = node_pool.get_node(start)
    while current:
        if current.state == goal:
            return reconstruct_path(current)

        for neighbor in get_neighbors(current):
            if neighbor not in visited:
                node = node_pool.get_node(neighbor)
                process_node(node, current, open_set)
```

8.5.2   Parallel Implementations

Parallel A* implementations can significantly improve performance on modern hardware.

Basic Parallel Structure

```
def parallel_a_star(start, goal, num_threads):
  shared_open_set = ConcurrentPriorityQueue()
  shared_closed_set = ConcurrentSet()
  def worker(thread_id):
      while not shared_open_set.empty():
          current = shared_open_set.get()


expand_node(current, shared_open_set, shared_closed_set)
  threads = [Thread(target=worker, args=(i,))
              for i in range(num_threads)]
```

Load Balancing

```
class ParallelSearchManager:
  def __init__(self, num_workers):
      self.workers = []
      self.work_queues = [Queue() for _ in range(num_workers)]

  def distribute_work(self, nodes):
      for i, node in enumerate(nodes):
          queue_index = i % len(self.work_queues)
          self.work_queues[queue_index].put(node)
```

8.5.3   Dynamic Environments Handling

Modern applications often require A* to work in changing environments.

Dynamic Cost Updates

```
class DynamicAstar:
  def __init__(self):
      self.cost_map = {}
      self.update_queue = Queue()

  def update_costs(self, changes):
      for location, new_cost in changes.items():
          self.cost_map[location] = new_cost
          affected_paths = self.find_affected_paths(location)
          self.replan_paths(affected_paths)
```

Incremental Search

```
def d_star_lite(start, goal, grid):
  def update_vertex(u):
      if u != goal:
          rhs = min(cost(u, s) + g[s] for s in succ(u))
          update_queue.insert(u, calculate_key(u))
```

8.5.4   Integration with Modern AI Systems

A* can be enhanced with modern AI techniques for improved performance.

Neural Network Heuristics

```
class NeuralHeuristic:
  def __init__(self, model_path):
      self.model = load_neural_network(model_path)

  def __call__(self, state):
      features = extract_features(state)
      return self.model.predict(features)
```

Reinforcement Learning Integration

```
class HybridPathfinder:
  def __init__(self, rl_model, traditional_planner):
      self.rl_model = rl_model
      self.planner = traditional_planner

  def find_path(self, start, goal):
      if self.is_simple_case(start, goal):
          return self.rl_model.get_path(start, goal)
      else:
          return self.planner.a_star(start, goal)
```

Performance Metrics

- Time efficiency: $O(b^d)$ theoretical, often better in practice;
- Memory usage: $O(b^d)$ worst case;
- Solution quality: Optimal with admissible heuristics.

These advanced techniques can significantly improve A*'s performance:

- Memory optimizations reduce space requirements;
- Parallel implementations speed up search;
- Dynamic handling enables real-time adaptation;
- AI integration improves heuristic accuracy.

Understanding these optimizations is crucial for implementing A* in modern, demanding applications. The choice of optimization technique depends on specific application requirements and constraints.

8.6   Practice and Exercises

8.6.1   Practice and Programming Exercises

All programming exercises for this chapter are available in the Google Colab notebook. You can access the complete notebook by scanning the QR code below or using this link: https://colab.research.google.com/drive/1ODWbRi8nzjbxToOLK3 ew4UKg2Ko1jye6

The notebook contains interactive Python code that you can run and modify to better understand A* algorithm. Each exercise builds upon the previous ones, gradually introducing more complex aspects of optimization algorithms. Solutions to the exercises are available to instructors through the publisher's resources.

Exercise 1: Basic A* Implementation (Notebook Cells 1–3)

Learning Objectives:
- Understand the core components of the A* algorithm;
- Implement the basic A* search function in Python;

- Learn to manage open and closed sets effectively;
- Master path reconstruction techniques.

Tasks:

1. Create a Node class with required attributes (state, g_score, h_score, parent);
2. Implement the main A* search function;
3. Write a path reconstruction function;
4. Test the implementation on a simple grid environment;
5. Visualize the search process and final path.

Exercise 2: Path Finding Analysis (Notebook Cells 4–6)

Learning Objectives:
- Compare different heuristic functions;
- Analyze algorithm performance;
- Understand the impact of terrain types on pathfinding;
- Master visualization techniques for search algorithms.

Tasks:

1. Implement Manhattan, Euclidean, and Diagonal distance heuristics;
2. Create a test environment with various terrain types;
3. Compare performance metrics across different heuristics;
4. Visualize search patterns and final paths;
5. Document and analyze the results.

Exercise 3: A* Variants Implementation (Notebook Cells 7–9)

Learning Objectives:
- Understand key differences between A* variants;
- Master memory-efficient pathfinding techniques;
- Learn to handle resource constraints;
- Develop skills in algorithm adaptation.

Tasks:

1. Implement IDA* (Iterative Deepening A*);
2. Create a memory-bounded version of A*;
3. Compare memory usage between variants;
4. Test variants on complex navigation scenarios;
5. Analyze trade-offs between time and space complexity.

Exercise 4: Performance Analysis (Notebook Cells 10–12)

Learning Objectives:
- Master algorithm performance measurement;
- Learn to optimize pathfinding implementations;
- Understand resource utilization patterns;
- Develop skills in comparative analysis.

Tasks:

1. Create comprehensive performance testing framework;
2. Measure and compare execution times;
3. Analyze memory consumption patterns;
4. Test scaling behavior with different map sizes;
5. Document optimization opportunities.

8.6.2   Self-Assessment Questions

Before attempting the questions:

1. Review the chapter material thoroughly;
2. Complete all practical exercises;
3. Understand the key concepts of A* and its variants.

Multiple Choice Questions

1. What is the primary characteristic that distinguishes A* from other search algorithms?

    (a) It always finds the shortest path
    (b) It combines actual path cost with heuristic estimation
    (c) It uses less memory than other algorithms
    (d) It runs faster than all other search algorithms

2. Which statement about A*'s heuristic function h(n) is correct for optimality?

    (a) It must exactly match the actual cost
    (b) It must never underestimate the actual cost
    (c) It must never overestimate the actual cost
    (d) It can estimate any value randomly

3. When implementing A*, what data structure is most appropriate for the open set?

    (a) Stack
    (b) Queue

  (c) Priority Queue
  (d) Array

4. What is the main advantage of IDA* over standard A*?

  (a) It always finds shorter paths
  (b) It runs faster
  (c) It uses less memory
  (d) It has better heuristics

5. Which of the following is NOT a valid admissible heuristic for grid-based pathfinding?

  (a) Manhattan distance
  (b) Euclidean distance
  (c) Sum of all grid costs
  (d) Diagonal distance

6. In the context of A*, what does "consistent heuristic" mean?

  (a) The heuristic never changes during search
  (b) The heuristic estimates are always correct
  (c) The heuristic satisfies the triangle inequality
  (d) The heuristic always returns the same value

7. What is the main purpose of SMA* (Simplified Memory-Bounded A*)?

  (a) To find paths faster than standard A*
  (b) To work within fixed memory constraints
  (c) To improve heuristic accuracy
  (d) To guarantee optimal paths

8. When is the Manhattan distance heuristic most appropriate?

  (a) In 3D environments
  (b) In grid-based environments with diagonal movement
  (c) In grid-based environments with only cardinal movements
  (d) In continuous space environments

9. What happens if A*'s heuristic function overestimates the actual cost?

  (a) The algorithm becomes faster
  (b) The algorithm may not find the optimal path
  (c) The algorithm uses less memory
  (d) The algorithm becomes more accurate

10. Which A* variant is most suitable for real-time applications?

  (a) Standard A*
  (b) IDA*
  (c) ARA* (Anytime Repairing A*)
  (d) SMA*

Answer Key and Explanations

1. (b) A* is unique in combining g(n) (actual path cost) with h(n) (heuristic estimate)
2. (c) For A* to guarantee optimality, h(n) must never overestimate the actual cost
3. (c) Priority queue ensures efficient selection of nodes with lowest f-value
4. (c) IDA*'s main advantage is its linear memory usage compared to A*'s exponential memory requirements
5. (c) Sum of all grid costs could overestimate the actual cost, violating admissibility
6. (c) A consistent heuristic satisfies the triangle inequality: $h(n) \leq c(n,n') + h(n')$
7. (b) SMA* was designed specifically to operate within fixed memory limits
8. (c) Manhattan distance is optimal for grid environments with only up/down/left/right movements
9. (b) Overestimation can lead to suboptimal paths as A* may ignore better alternatives
10. (c) ARA* is designed for real-time applications by quickly finding initial solutions and improving them over time

8.7   Conclusion

The A* algorithm represents a powerful tool in modern artificial intelligence. This chapter has presented its theoretical foundations, practical implementations, and real-world applications. The progression from basic concepts to advanced techniques provides comprehensive understanding of this fundamental algorithm.

Our experimental results demonstrate A*'s versatility across different scenarios. The comparison of various heuristics and algorithm variants reveals important performance trade-offs. These findings inform implementation choices in practical applications.

Key insights from our practical exercises include:

1. Different heuristic functions often achieve similar results despite theoretical distinctions.
2. Algorithm variant selection depends primarily on application requirements.
3. Memory constraints and execution time priorities significantly impact implementation choices.
4. Implementation simplicity often outweighs marginal performance gains.

These insights help students select and implement the appropriate algorithm variant for specific scenarios. The balance between theoretical understanding and practical application prepares students to apply A* effectively in diverse problem domains.

References

1. Hart, P.E., Nilsson, N.J., Raphael, B.: A Formal Basis for the Heuristic Determination of Minimum Cost Paths. IEEE Transactions on Systems Science and Cybernetics. 4, 100–107 (1968). https://doi.org/10.1109/TSSC.1968.300136.
2. Dijkstra, E.W.: A note on two problems in connexion with graphs. Numer. Math. 1, 269–271 (1959). https://doi.org/10.1007/BF01386390.
3. Frana, P.L., Misa, T.J.: An interview with Edsger W. Dijkstra. Commun. ACM. 53, 41–47 (2010). https://doi.org/10.1145/1787234.1787249.
4. Norvig, P.: Artificial intelligence: a modern approach. Global edition. Pearson, Boston (2021).
5. Delling, D., Sanders, P., Schultes, D., Wagner, D.: Engineering Route Planning Algorithms. In: Lerner, J., Wagner, D., and Zweig, K.A. (eds.) Algorithmics of Large and Complex Networks: Design, Analysis, and Simulation. pp. 117–139. Springer, Berlin, Heidelberg (2009). https://doi.org/10.1007/978-3-642-02094-0_7.
6. Zeng, W., Church, R.L.: Finding shortest paths on real road networks: the case for A*. International Journal of Geographical Information Science. 23, 531–543 (2009). https://doi.org/10.1080/13658810801949850.
7. Xie, J., Wang, N., Huang, X., Lei, Q., Molins, C.: A novel A-star algorithm-based approach to predicting the sealant performance of shield tunnel's gasketed joints. Tunnelling and Underground Space Technology. 158, 106379 (2025). https://doi.org/10.1016/j.tust.2025.106379.
8. Huang, J., Chen, C., Shen, J., Liu, G., Xu, F.: A self-adaptive neighborhood search A-star algorithm for mobile robots global path planning. Computers and Electrical Engineering. 123, 110018 (2025). https://doi.org/10.1016/j.compeleceng.2024.110018.
9. Zuo, S., Mao, Z., Fan, C., Chen, X., Gong, M., Ren, J., Fan, X., Guo, Y.: Dynamic planning of crowd evacuation path for metro station based on Dynamic Avoid Smoke A-Star algorithm. Tunnelling and Underground Space Technology. 154, 106145 (2024). https://doi.org/10.1016/j.tust.2024.106145.
10. Miyombo, M.E., Liu, Y., Mulenga, C.M., Siamulonga, A., Kabanda, M.C., Shaba, P., Xi, C., Ayodeji, A.: Optimal path planning in a real-world radioactive environment: A comparative study of A-star and Dijkstra algorithms. Nuclear Engineering and Design. 420, 113039 (2024). https://doi.org/10.1016/j.nucengdes.2024.113039.
11. Erdoğan, H.: Entomopathogenic nematode detection and counting model developed based on A-star algorithm. Journal of Invertebrate Pathology. 207, 108196 (2024). https://doi.org/10.1016/j.jip.2024.108196.

Chapter 9
Genetic Algorithms

Abstract This chapter explores genetic algorithms as optimization tools inspired by natural evolution. We present both theoretical principles and practical implementations. The material progresses from basic concepts to advanced techniques. Students learn chromosome representation, fitness evaluation, selection mechanisms, and genetic operators. Python examples demonstrate implementation strategies for various problems. Real-world applications include portfolio optimization and constraint handling. Performance analysis helps students understand algorithm behavior.

9.1 Foundations of Genetic Algorithms

Genetic algorithms (GAs) represent a fascinating intersection of biology and computer science [1, 2]. They draw inspiration from natural evolution to solve complex optimization problems [3, 4]. This section explores their foundations, principles, and relationship to other optimization methods.

9.1.1 Historical Context and Biological Inspiration

John Holland introduced genetic algorithms in 1975 through his groundbreaking book "Adaptation in Natural and Artificial Systems" [5]. His work bridged the gap between biological evolution and computational problem-solving [6]. The fundamental insight was that the principles of natural selection could be adapted to solve complex computational problems.

The biological inspiration comes from several key evolutionary processes [5]:

- Natural Selection: In nature, organisms better adapted to their environment have a higher chance of survival and reproduction. In GAs, solutions better suited to solving the problem have a higher chance of being selected for reproduction.

O. Kuznetsov, *Intelligent Systems: From Theory to Applications*, Cognitive Technologies, https://doi.org/10.1007/978-3-032-00044-6_9

- Inheritance: Just as biological offspring inherit traits from their parents through genes, new solutions in GAs inherit characteristics from their parent solutions through a process called crossover.
- Mutation: Random genetic changes occur in nature, introducing variation into populations. Similarly, GAs use mutation operators to maintain diversity in the solution population.

9.1.2 Basic Principles and Terminology

The GA terminology directly parallels biological concepts [3, 6]:

- Chromosome: A candidate solution to the problem, typically encoded as a string of values. For example, in binary encoding, a chromosome might look like: $x = (1, 0, 1, 1, 0, 1)$.
- Gene: A single element of the chromosome representing one aspect of the solution. In the binary example above, each 0 or 1 is a gene.
- Population: A set of chromosomes (candidate solutions): $P = x_1, x_2, \ldots, x_n$.
- Fitness Function: A measure $f(x)$ that evaluates how good a solution is. Higher fitness values indicate better solutions.
- Generation: One complete cycle of the genetic algorithm, including selection, reproduction, and replacement of solutions.

The terminology and concepts introduced above form the foundation for understanding genetic algorithms. These biological analogies provide an intuitive framework for grasping how GAs solve complex optimization problems. While the terminology may seem abstract at first, it directly maps to practical implementation components we will explore in later sections.

9.1.3 Components of Genetic Algorithms

A genetic algorithm consists of five essential components [5, 6]:

1. Representation: The way solutions are encoded. The formal definition is: $x \in {0, 1}^n$ for binary encoding and $x \in \mathrm{R}^n$ for real-valued encoding.
2. Fitness Function: A mapping from solutions to quality measures: $f : X \to \mathrm{R}$.
3. Selection Operator: A mechanism to choose parents for reproduction: $s : P \to P$.
4. Genetic Operators:

   - Crossover: $c : X \times X \to X \times X$.
   - Mutation: $m : X \to X$.

5. Replacement Strategy: Determines how new solutions replace old ones in the population.

9.1.4   Relationship to Other Optimization Methods

Genetic algorithms differ from traditional optimization methods in several key ways:

- Population-Based: Unlike gradient descent or simulated annealing, GAs work with multiple solutions simultaneously. This parallel search helps avoid local optima.
- Stochastic Nature: GAs use probabilistic transitions, making them more robust in noisy or discontinuous search spaces.
- No Gradient Information: GAs don't require gradient information, making them suitable for non-differentiable or discrete problems.

Table 9.1 contrasts key characteristics of different optimization approaches, highlighting the unique advantages of genetic algorithms in handling multiple solutions and operating without gradient information.

This foundation of genetic algorithms provides the framework for understanding their implementation and application, which we will explore in subsequent sections. The combination of biological inspiration and computational efficiency makes GAs particularly effective for complex optimization problems where traditional methods may struggle [5].

Having established the theoretical foundations, we now turn to the practical aspects of implementing genetic algorithms. Each component discussed below plays a crucial role in the algorithm's effectiveness. Understanding these components and their interactions is essential for successful GA implementation.

9.2   Core Components and Operators

This section explores the fundamental components and operators that make genetic algorithms work. Understanding these elements is crucial for implementing effective genetic algorithms.

The core components of genetic algorithms work together to simulate evolutionary processes. Each component serves a specific purpose in the optimization process: representation determines how solutions are encoded, fitness functions evaluate solution quality, selection drives evolutionary pressure, and genetic operators create new candidate solutions. Understanding how these components interact is crucial for effective implementation.

Table 9.1 Comparison of optimization methods

Method	Search strategy	Gradient required	Multiple solutions
Gradient descent	Local	Yes	No
Simulated annealing	Local with jumps	No	No
Genetic algorithms	Global	No	Yes

9.2.1 Population Representation and Initialization

Solution representation is perhaps the most critical design decision when implementing a genetic algorithm. The choice of representation affects all other components and directly impacts algorithm performance. While several standard representations exist, each has specific advantages and limitations [6]:

- Binary Representation: The simplest and most traditional encoding where solutions are represented as binary strings: $x = (b_1, b_2, \ldots, b_n)$, where $b_i \in 0, 1$.
- Integer Representation: Useful for discrete optimization problems: $x = (i_1, i_2, \ldots, i_n)$, where $i_j \in Z$.
- Real-Value Representation: Appropriate for continuous optimization problems: $x = (r_1, r_2, \ldots, r_n)$, where $r_k \in R$.
- Permutation Representation: Essential for ordering problems like the traveling salesman problem: $x = (p_1, p_2, \ldots, p_n)$, where p_m is a permutation of 1, 2, …, n.

Population initialization typically uses random generation within valid ranges. A population of size N is represented as: $P = x_1, x_2, \ldots, x_N$.

9.2.2 Fitness Function Design

The fitness function $f(x)$ quantifies solution quality. Good fitness functions should [6]:

- Reflect the optimization objectives.
- Handle constraints appropriately.
- Scale well across the solution space.

For a minimization problem with constraints, a typical fitness function might be:

$$f(x) = g(x) + \sum_{i=1}^{m} p_i(x),$$

where:

- $g(x)$ is the objective function;
- $p_i(x)$ are penalty terms for constraint violations.

The design of an effective fitness function requires careful consideration of both the optimization objectives and problem constraints. A well-designed fitness function should provide meaningful differentiation between solutions while guiding the search toward feasible regions of the solution space. The penalty function approach shown above helps balance these requirements by incorporating constraint violations into the fitness calculation.

9.2.3   Selection Mechanisms

Selection operators choose parents for reproduction.
Common methods include [6]:

Roulette Wheel Selection

Selection probability proportional to fitness:

$$P(x_i) = \frac{f(x_i)}{\sum_{j=1}^{N} f(x_j)}.$$

Tournament Selection

Choose the best from k randomly selected individuals. Selection pressure can be adjusted by changing k.

Rank Selection

Selection probability based on sorted position:

$$P(x_i) = \frac{2r_i}{N(N+1)},$$

where r_i is the rank of individual i.

9.2.4   Crossover Operators

Crossover combines parent solutions to create offspring.
Key variants include [6]:

Single-Point Crossover

For parents x_1 and x_2, with crossover point k:

$$\left(x_{11},\ldots,x_{1k},x_{1,k+1},\ldots,x_{1n} \right) + \left(x_{21},\ldots,x_{2k},x_{2,k+1},\ldots,x_{2n} \right) \downarrow$$
$$\left(x_{11},\ldots,x_{1k},x_{2,k+1},\ldots,x_{2n} \right) + \left(x_{21},\ldots,x_{2k},x_{1,k+1},\ldots,x_{1n} \right)$$

Multi-Point Crossover

Exchanges multiple segments between parents.

Uniform Crossover

Each gene comes from either parent with probability 0.5.

9.2.5   *Mutation Operators*

Mutation introduces small random changes to maintain diversity:

- Bit-Flip Mutation (for binary representation): Each bit flips with probability p_m.
- Gaussian Mutation (for real values): $x_i' = x_i + N(0,\sigma)$.
- Swap Mutation (for permutations): Exchanges positions of two randomly chosen elements.

9.2.6   *Replacement Strategies*

Replacement determines how new solutions enter the population:

- Generational Replacement: The entire population is replaced each generation.
- Steady-State Replacement: Only a few individuals are replaced at a time.
- Elitism: The best e individuals always survive to the next generation. This ensures the best solutions are not lost.

The combination of these operators must be carefully balanced. Selection pressure should promote good solutions while maintaining diversity. Crossover should effectively combine good solution components. Mutation should prevent premature convergence without disrupting too much progress.

A typical evolutionary cycle can be expressed as:

$$P_{t+1} = r\big(s\big(c\big(m(P_t)\big)\big)\big),$$

where:

- P_t is the population at generation t;
- m is mutation;
- c is crossover;
- s is selection;
- r is replacement.

The effectiveness of genetic algorithms depends heavily on choosing appropriate operators and parameters for the specific problem at hand.

9.3   Implementation Details

Moving from theoretical understanding to practical implementation requires careful attention to numerous implementation details. This section bridges theory and practice by providing concrete Python implementations and discussing crucial implementation considerations. Each code example demonstrates key concepts while highlighting important design decisions.

9.3.1   Population Management

Population size affects both solution quality and computational cost. A typical range is: $50 \leq N \leq 200$.

Here's a basic Python implementation for population initialization:

```python
def initialize_population(pop_size, chromosome_length):
  return np.random.randint(2, size=(pop_size, chromosome_length))
```

Population diversity should be monitored using metrics like:

$$D(P) = \frac{1}{N} \sum_{i=1}^{N} \sum_{j=i+1}^{N} d\left(x_i, x_j\right),$$

where $d(x_i, x_j)$ is the distance between two individuals.

9.3.2   Parameter Settings and Tuning

Key parameters require careful tuning:

- Crossover Rate (p_c). Typical range: $0.6 \leq p_c \leq 0.9$

- Mutation Rate (p_m). Common setting: $p_m = \frac{1}{l}$ where l is chromosome length

- Population Size (N). Scales with problem complexity: $N = k\sqrt{l}$ where k is a constant (typically 5–10)

Here's a Python class implementing these parameters:

```
class GeneticAlgorithm:
  def __init__(self, pop_size=100, p_c=0.8, p_m=0.01):
      self.pop_size = pop_size
      self.p_c = p_c
      self.p_m = p_m
```

9.3.3   Convergence Criteria

Multiple termination conditions should be implemented:

- Maximum Generations:

```
if generation >= max_generations:
  break
```

- Fitness Convergence:

```
if abs(best_fitness - prev_best) < epsilon:
  stall_count += 1
if stall_count > max_stall:
  break
```

- Diversity Check: $D(P) < D_{threshold}$.

9.3.4   Handling Constraints

Constraints can be handled through:

Penalty Functions

$$f_{penalized}(x) = f(x) + \sum_{i=1}^{m} w_i \max\left(0, g_i(x)\right).$$

```
def evaluate_with_penalties(solution, weights):
  fitness = calculate_fitness(solution)
  penalties = sum(w * max(0, g(solution))
               for w, g in zip(weights, constraints))
  return fitness + penalties
```

Repair Mechanisms

```python
def repair_solution(solution):
  while not is_feasible(solution):
      solution = adjust_solution(solution)
  return solution
```

9.3.5   Basic Python Implementation

Here's a complete basic implementation:

```python
class GeneticAlgorithm:
  def __init__(self, fitness_func, chromosome_length,
               pop_size=100, p_c=0.8, p_m=0.01):
      self.fitness_func = fitness_func
      self.pop_size = pop_size
      self.chromosome_length = chromosome_length
      self.p_c = p_c
      self.p_m = p_m

  def evolve(self, generations=100):
      population = self.initialize_population()
      for gen in range(generations):
          fitness_values = self.evaluate_population(population)
          parents = self.select_parents(population,
fitness_values)
          offspring = self.crossover(parents)
          offspring = self.mutate(offspring)
          population = self.replacement(population, offspring)

          if self.check_convergence():
              break

      return self.get_best_solution(population)
```

9.3.6   Best Practices

1. Code Organization:

   - Separate genetic operators into distinct functions.
   - Use object-oriented design for maintainability.
   - Implement logging and visualization tools.

2. Performance Optimization:

   - Vectorize operations using NumPy.
   - Cache fitness values.
   - Implement parallel evaluation when possible.

3. Robustness:

```python
def safe_genetic_operation(func):
  def wrapper(*args, **kwargs):
      try:
          return func(*args, **kwargs)
      except Exception as e:
          logging.error(f"Error in {func.__name__}: {e}")
          return fallback_operation(*args, **kwargs)
  return wrapper
```

4. Testing and Validation:

   - Use benchmark problems.
   - Implement unit tests for each operator.
   - Monitor population statistics.

Implementation success often depends on careful attention to these details. The code should be both efficient and maintainable, with clear documentation and robust error handling.

9.4 Advanced Topics

This section explores advanced concepts and techniques that extend the basic genetic algorithm framework. These topics are essential for tackling complex real-world optimization problems. These advanced concepts extend the basic genetic algorithm framework to handle more complex real-world optimization scenarios. While the basic GA provides good results for many problems, these enhancements can significantly improve performance in challenging optimization landscapes. Each advanced technique addresses specific limitations of the basic algorithm.

9.4.1 *Multi-Objective Optimization*

Many real-world problems involve multiple competing objectives. A multi-objective genetic algorithm (MOGA) seeks to find Pareto-optimal solutions [7].

The formal definition of a multi-objective optimization problem is:

- Minimize: $F(x) = (f_1(x), f_2(x), \ldots, f_k(x))$.
- Subject to: $g_i(x) \leq 0$, $i = 1, \ldots, m$ $h_j(x) = 0$, $j = 1, \ldots, p$.

A solution x_1 dominates x_2 if:

- $\forall i : f_i(x_1) \leq f_i(x_2)$.
- $\exists j : f_j(x_1) < f_j(x_2)$.

The NSGA-II algorithm is a popular implementation [8, 9]:

```python
def nsga2_sort(population):
    fronts = []
    dominated_count = {}
    dominated_solutions = {}
    for p in population:
        dominated_solutions[p] = []
        dominated_count[p] = 0

        for q in population:
            if dominates(p, q):
                dominated_solutions[p].append(q)
            elif dominates(q, p):
                dominated_count[p] += 1

    return non_dominated_sort(dominated_count, dominated_solutions)
```

9.4.2 Adaptive Parameter Control

Static parameters often limit GA performance. Adaptive control adjusts parameters during evolution [10, 11].

Learning Rate Adaptation

$$p_m(t+1) = p_m(t) \cdot (1 + \alpha \cdot \Delta f),$$

where:

- $p_m(t)$ is the mutation rate at generation t;
- Δf is the relative fitness improvement;
- α is a learning rate.

Population Size Adaptation

$$N(t+1) = N(t) \cdot \left(1 + \beta \cdot D(P)\right),$$

where:

- $N(t)$ is the population size at generation t;
- $D(P)$ is the population diversity;
- β is a scaling factor.

9.4.3 Parallel Implementation

Parallel GAs can significantly reduce computation time [12, 13]. Three main approaches exist:

Global Single-Population

```
def parallel_evaluation(population):
  with Pool(processes=n_cores) as pool:
      fitness_values = pool.map(evaluate_fitness, population)
  return fitness_values
```

Island Model

Multiple sub-populations evolve independently with occasional migration: $P = P_1$, $P_2, \ldots, P_k$.

Migration Occurs Every m Generations

```
def migrate(islands, migration_rate):
  for i in range(len(islands)):
      for j in range(len(islands)):
          if i != j:
              exchange_individuals(islands[i], islands[j],
                                   migration_rate)
```

9.4.4 Hybrid Approaches

Hybrid algorithms combine GAs with other optimization methods [14, 15].

Memetic Algorithms

Combine genetic search with local optimization:

```python
def memetic_optimization(population):
  for individual in population:
      improved = local_search(individual)
      if fitness(improved) > fitness(individual):
          individual = improved
  return population
```

Lamarckian Evolution

Local improvements are inherited: $x'inherited = local_{search}(xparent)$.

9.4.5 *Performance Enhancement Techniques*

Several techniques can improve GA performance:

Fitness Caching

```python
class CachedFitness:
  def __init__(self):
      self.cache = {}

  def evaluate(self, solution):
      key = hash(solution.tobytes())
      if key not in self.cache:
          self.cache[key] = compute_fitness(solution)
      return self.cache[key]
```

Dynamic Scaling

Adjust fitness values to maintain selection pressure:

$$f'(x) = a \cdot f(x) + b,$$

where a and b are chosen to maintain desired selection pressure.

Niching

Maintain population diversity through fitness sharing:

$$f'(x_i) = \frac{f(x_i)}{\sum_{j=1}^{N} sh\left(d\left(x_i, x_j\right)\right)},$$

where $sh(d)$ is the sharing function:

$$sh(d) = \max\left(0, 1 - \left(\frac{d}{\sigma_{share}}\right)^{\alpha}\right).$$

These advanced techniques can significantly improve GA performance on complex problems. However, they often require careful tuning and problem-specific adaptations. The choice of which techniques to use should be guided by the specific characteristics of the optimization problem and available computational resources.

9.5   Practical Applications

This section explores real applications of genetic algorithms through classical problems and case studies. Understanding these applications helps bridge the gap between theory and practice.

The true value of genetic algorithms becomes apparent when applying them to real-world optimization problems. These examples demonstrate how theoretical concepts translate into practical solutions. Each application highlights different aspects of GA implementation and shows how to adapt the algorithm to specific domain requirements.

9.5.1   Classical Optimization Problems

The Traveling Salesman Problem (TSP)

Given n cities and distances between them, find the shortest route visiting each city exactly once.

Mathematical formulation. Minimize:

$$\sum_{i=1}^{n-1} d\left(c_i, c_{i+1}\right) + d\left(c_n, c_1\right),$$

where $d(c_i, c_j)$ is the distance between cities i and j.

Implementation example:

```python
def tsp_fitness(route):
    total_distance = 0
    for i in range(len(route)-1):
        total_distance += distance_matrix[route[i]][route[i+1]]
    total_distance += distance_matrix[route[-1]][route[0]]
    return -total_distance  # Negative because we minimize distance
```

The Knapsack Problem

Select items to maximize value while respecting weight constraints.
  Formulation. Maximize:

$$\sum_{i=1}^{n} v_i x_i,$$

subject to:

$$\sum_{i=1}^{n} w_i x_i \leq W,$$

where: $x_i \in 0, 1$

9.5.2   Real-World Case Studies

Circuit Design Optimization

Goal: Minimize circuit area while meeting performance requirements.
  Chromosome: Component placement and routing.
  Fitness:

$$f(x) = -\left(\text{area} + \alpha \cdot \text{delay} + \beta \cdot \text{power}\right).$$

```python
class CircuitOptimizer:
    def evaluate_circuit(self, layout):
        area = calculate_area(layout)
        delay = simulate_delay(layout)
        power = estimate_power(layout)
        return -(area + self.alpha * delay + self.beta * power)
```

Financial Portfolio Optimization

Goal: Maximize return while minimizing risk.
   Chromosome: Asset weights.
   Fitness:

$$f(w) = E\left[R_p\right] - \lambda \sigma_p^2,$$

where:

- $E[R_p]$ is expected portfolio return;
- σ_p^2 is portfolio variance;
- λ is risk aversion parameter.

9.5.3   Implementation Considerations

Problem-Specific Encoding: Choose representation that preserves solution validity.

```
def encode_portfolio(weights):
  # Ensure weights sum to 1
  return weights / np.sum(weights)
```

Constraint Handling: Use repair mechanisms or penalty functions.

```
def repair_solution(solution):
  if sum(solution) > max_weight:
      return normalize_to_constraint(solution)
  return solution
```

Performance Optimization: Cache expensive computations.

```
@lru_cache(maxsize=1000)
def expensive_fitness_calculation(solution):
  return complex_evaluation(solution)
```

9.5.4   Common Challenges and Solutions

Premature Convergence

Challenge: Population loses diversity too quickly.
   Solution: Implement niching or island models.

```python
def calculate_niche_count(population):
  niche_counts = np.zeros(len(population))
  for i in range(len(population)):
      for j in range(len(population)):
          distance = hamming_distance(population[i],
population[j])
          if distance < niche_radius:
              niche_counts[i] += 1
  return niche_counts
```

Handling Noisy Fitness

Challenge: Fitness evaluation contains noise.
  Solution: Multiple evaluations or statistical measures.

```python
def robust_fitness(solution, n_samples=10):
  evaluations = [evaluate(solution) for _ in range(n_samples)]
  return statistics.mean(evaluations)
```

Computational Cost

Challenge: Slow fitness evaluation.
  Solution: Parallel evaluation and surrogate models.

```python
def parallel_fitness_evaluation(population):
  with ProcessPoolExecutor() as executor:
      return list(executor.map(fitness_function, population))
```

Scale-Up Issues

Challenge: Performance degrades with problem size.
  Solution: Hierarchical decomposition.

```python
def hierarchical_optimization(problem):
  sub_problems = decompose_problem(problem)
  solutions = [optimize_sub_problem(p) for p in sub_problems]
  return combine_solutions(solutions)
```

These practical applications demonstrate how genetic algorithms can be adapted to solve diverse real-world problems. Success often depends on careful problem formulation and appropriate handling of domain-specific constraints and challenges.

9.6 Practice and Exercises

All programming exercises for this chapter are available in the Google Colab note-book. You can access the complete notebook by scanning the QR code below or using this link: https://colab.research.google.com/drive/1hZoJu9Kv5qwsSGFye2A nuG64sHNCzuM0

The notebook contains interactive Python code that you can run and modify to better understand genetic algorithms. Each exercise builds upon the previous ones, gradually introducing more complex aspects of genetic algorithm implementation. Solutions to the exercises are available to instructors through the publisher's resources.

9.6.1 Practice and Programming Exercises

Exercise 1: Basic Genetic Algorithm (Cells 1–3)

Implement a basic genetic algorithm for the knapsack problem. This exercise demonstrates fundamental concepts through a simple implementation.

Learning Objectives:
- Understand fundamental genetic algorithm components;
- Learn basic population representation;
- Master simple genetic operators;
- Grasp fitness function design.

Tasks:
- Implement basic population representation for knapsack problem;
- Create fitness function incorporating weight constraints;
- Develop simple selection mechanism (roulette wheel);
- Implement single-point crossover operator;

- Add basic mutation operator;
- Visualize fitness evolution.

Exercise 2: Enhanced Genetic Algorithm (Cells 4–6)

Extend the basic implementation with improved operators and monitoring capabilities.

Learning Objectives:
- Master advanced selection methods;
- Understand population diversity maintenance;
- Learn constraint handling techniques;
- Develop performance monitoring skills.

Tasks:
- Implement tournament selection;
- Add multi-point crossover operator;
- Create adaptive mutation rate mechanism;
- Develop constraint repair functions;
- Add population diversity tracking;
- Visualize multiple performance metrics.

Exercise 3: Advanced Genetic Algorithm (Cells 7–9)

Implement an advanced genetic algorithm with adaptive parameters and comprehensive monitoring.

Learning Objectives:
- Master parameter adaptation techniques;
- Understand elitism implementation;
- Learn comprehensive monitoring;
- Grasp advanced optimization concepts.

Tasks:
- Implement parameter self-adaptation;
- Add elitism mechanism;
- Create detailed performance tracking;
- Develop multiple genetic operators;
- Add constraint violation monitoring;
- Create comprehensive visualization system.

Exercise 4: Performance Analysis (Cells 10)

Compare different genetic algorithm variants across multiple problem instances.

Learning Objectives:
- Understand algorithm comparison methods;
- Learn performance metric design;
- Master scalability analysis;
- Grasp statistical evaluation techniques.

Tasks:

- Implement benchmark problem generator;
- Create performance comparison framework;
- Analyze scalability with problem size;
- Compare different correlation types;
- Evaluate statistical significance;
- Visualize comparative results.

Exercise 5: Portfolio Optimization (Cells 11–14)

Apply genetic algorithms to real-world financial optimization.

Learning Objectives:
- Apply GAs to real-world problems;
- Understand financial constraints;
- Learn risk management techniques;
- Master multi-objective optimization.

Tasks:
- Implement portfolio representation;
- Create risk-return fitness function;
- Add investment constraints;
- Develop risk management mechanisms;
- Create portfolio rebalancing logic;
- Visualize portfolio allocation results.

9.6.2   Self-Assessment Questions

Before attempting the questions:

- Review the chapter material thoroughly;
- Complete all practical exercises;
- Understand both theoretical concepts and their practical applications.

Multiple Choice Questions

1. What is the primary advantage of using crossover in genetic algorithms?

   (a) It increases mutation rate
   (b) It combines good features from different solutions
   (c) It reduces population size
   (d) It speeds up convergence by removing poor solutions

2. When implementing a genetic algorithm for the knapsack problem, what is the most appropriate way to handle constraint violations?

   (a) Reject all invalid solutions immediately
   (b) Use penalty functions in fitness calculation
   (c) Ignore constraints during evolution
   (d) Reduce population size

3. In genetic algorithms, what does elitism ensure?

   (a) Faster convergence to local optima
   (b) Higher mutation rates
   (c) Preservation of the best solutions
   (d) Maximum population diversity

4. Which parameter has the most significant impact on genetic algorithm performance?

   (a) Population size
   (b) Number of generations
   (c) Crossover type
   (d) Selection pressure

5. When should adaptive mutation rates be used in genetic algorithms?

   (a) Only for small populations
   (b) When fitness stagnates
   (c) In every generation
   (d) For initial population only

6. What is the main purpose of tournament selection?

   (a) To maintain population diversity
   (b) To speed up convergence
   (c) To control selection pressure
   (d) To reduce computational cost

7. In portfolio optimization using genetic algorithms, what is the primary constraint?

   (a) Maximum return
   (b) Portfolio weights sum to 1

   (c) Minimum risk
   (d) Number of assets

8. How can premature convergence be prevented in genetic algorithms?

   (a) Increase population size
   (b) Reduce crossover rate
   (c) Maintain population diversity
   (d) Use only mutation

9. What does a fitness landscape represent in genetic algorithms?

   (a) Population distribution
   (b) Solution quality mapping
   (c) Mutation probability
   (d) Selection pressure

10. Which metric best indicates genetic algorithm performance?

   (a) Final population size
   (b) Convergence speed
   (c) Best fitness achieved
   (d) Number of generations

Answers

1. (b) Crossover combines beneficial features from parent solutions.
2. (b) Penalty functions help guide search while maintaining diversity.
3. (c) Elitism ensures best solutions survive between generations.
4. (a) Population size affects exploration-exploitation balance.
5. (b) Adaptive mutation helps escape local optima when progress stalls.
6. (c) Tournament selection allows direct control of selection pressure.
7. (b) Portfolio weights must sum to 1 for valid allocation.
8. (c) Maintaining diversity prevents early convergence to suboptimal solutions.
9. (b) Fitness landscape shows quality distribution across solution space.
10. (c) Best fitness indicates solution quality achievement.

9.7 Conclusions and Future Directions

Genetic algorithms offer powerful approaches to complex optimization challenges.
This chapter has shown how evolutionary principles translate into effective computational methods. The population-based search strategy provides significant advantages for problems with large solution spaces, multiple optima, or complex
constraints.

Key insights from our exploration include the critical importance of representation choice, fitness function design, and operator selection. These decisions significantly impact algorithm performance. The practical examples, particularly portfolio optimization, demonstrate how these theoretical concepts solve real problems effectively.

Recent developments point to several promising directions. Adaptive parameter control reduces the need for manual tuning. Parallel implementations leverage modern computing architectures. Hybrid approaches combine genetic algorithms with other optimization techniques. Multi-objective methods address problems with competing goals.

References

1. Martin, W.N., Spears, W.M.: Introduction. In: Martin, W.N. and Spears, W.M. (eds.) Foundations of Genetic Algorithms 6. pp. 1–3. Morgan Kaufmann, San Francisco (2001). https://doi.org/10.1016/B978-155860734-7/50083-4.
2. Cox, E.: Introduction. In: Cox, E. (ed.) Fuzzy Modeling and Genetic Algorithms for Data Mining and Exploration. pp. xix–xxi. Morgan Kaufmann, San Francisco (2005). https://doi.org/10.1016/B978-012194275-5/50002-5.
3. D'Ambrosio, D., Spataro, W., Rongo, R., Iovine, G.G.R.: 2.7 Genetic Algorithms, Optimization, and Evolutionary Modeling. In: Shroder, J.F. (ed.) Treatise on Geomorphology. pp. 74–97. Academic Press, San Diego (2013). https://doi.org/10.1016/B978-0-12-374739-6.00033-6.
4. Leardi, R.: 1.20—Genetic Algorithms. In: Brown, S.D., Tauler, R., and Walczak, B. (eds.) Comprehensive Chemometrics. pp. 631–653. Elsevier, Oxford (2009). https://doi.org/10.1016/B978-044452701-1.00039-9.
5. Holland, J.H.: Adaptation in Natural and Artificial Systems: An Introductory Analysis with Applications to Biology, Control, and Artificial Intelligence. University of Michigan Press (1975).
6. Petrowski, A., Ben-Hamida, S.: Evolutionary Algorithms. John Wiley & Sons (2017).
7. Imani, V., Sevilla-Salcedo, C., Moradi, E., Fortino, V., Tohka, J.: Multi-objective genetic algorithm for multi-view feature selection. Applied Soft Computing. 167, 112332 (2024). https://doi.org/10.1016/j.asoc.2024.112332.
8. Jiang, Q., Wang, P.: NSGA-II algorithm based control parameters optimization strategy for megawatt novel nuclear power systems. Energy. 316, 134444 (2025). https://doi.org/10.1016/j.energy.2025.134444.
9. Luo, J., Gu, Q., Chen, L., Li, X., Li, P.: Multi-objective optimization for ore blending schemes in the open-pit phosphate mine using an improved NSGA-II algorithm. Green and Smart Mining Engineering. (2025). https://doi.org/10.1016/j.gsme.2024.12.004.
10. Bakdi, A., Hentout, A., Boutami, H., Maoudj, A., Hachour, O., Bouzouia, B.: Optimal path planning and execution for mobile robots using genetic algorithm and adaptive fuzzy-logic control. Robotics and Autonomous Systems. 89, 95–109 (2017). https://doi.org/10.1016/j.robot.2016.12.008.
11. Abdul-Rahman, O.A., Munetomo, M., Akama, K.: An adaptive parameter binary-real coded genetic algorithm for constraint optimization problems: Performance analysis and estimation of optimal control parameters. Information Sciences. 233, 54–86 (2013). https://doi.org/10.1016/j.ins.2013.01.005.
12. Sharma, H., Galván, E., Mooney, P.: A parallel genetic algorithm for multi-criteria path routing on complex real-world road networks. Applied Soft Computing. 170, 112559 (2025). https://doi.org/10.1016/j.asoc.2024.112559.

13. Lu, W., Gao, W., Liu, B., Niu, W., Peng, X., Yang, Z., Song, Y.: Parallel dual adaptive genetic algorithm: A method for satellite constellation task assignment in time-sensitive target tracking. Advances in Space Research. 74, 5192–5213 (2024). https://doi.org/10.1016/j.asr.2024.07.044.
14. Banzhaf, W., Hu, T.: Evolutionary Computation. In: Evolutionary Biology. Oxford University Press (2019). https://doi.org/10.1093/obo/9780199941728-0122.
15. Norvig, P.: Artificial intelligence: a modern approach. Global edition. Pearson, Boston (2021).

Chapter 10
Hill Climbing

Abstract This chapter examines Hill Climbing, a fundamental optimization technique in artificial intelligence. We present both theoretical foundations and algorithm implementations. Several variants receive detailed treatment, including Steepest Ascent, Stochastic, and Random-Restart approaches. The material highlights each method's strengths and limitations. Python implementations demonstrate practical applications to various problems.

10.1 Introduction to Hill Climbing

Hill Climbing is a simple and intuitive one of the fundamental artificial intelligence algorithms for optimization. In this chapter, we introduce the fundamentals of Hill Climbing and illustrate its application in current AI frameworks.

10.1.1 Basic Principles and Motivation

Hill Climbing is guided by a simple but powerful principle [1, 2]: iteratively moving towards an improvement in a sequence of neighboring states. The algorithm name originates in an analogy with real-life hill climbing: a climber moving in an upward direction at all times in an attempt to climb to a peak.

The simple algorithm can be represented mathematically as [3, 4]:

1. Begin with an initial solution s
2. Examine neighbors of s in the solution space
3. Move to neighboring s' with best value, if it is an improving move
4. Repeat until a better neighbor

213

O. Kuznetsov, *Intelligent Systems: From Theory to Applications*, Cognitive
Technologies, https://doi.org/10.1007/978-3-032-00044-6_10

This process is captured in the following iterative formula:

$$s_{t+1} = \text{argmax}_{s' \in N(s_t)} f\left(s'\right)$$

where:

- s_t is the current solution at time t
- $N(s_t)$ is the neighborhood of s_t
- $f(s)$ is the objective function to be maximized

10.1.2   Historical Context

Hill Climbing was first used in early artificial intelligence research in the 1950s and evolved significantly over the years. It was originally a simple version of an optimization algorithm, and its derivation is from a metaphorical perspective of a climber looking for a peak, always moving upwards. This intuitive approach made it one of the first algorithmic solutions for optimization problems in early AI systems.

In the 1960s and 1970s, scientists started investigating several adaptations to enhance the simple algorithm [3, 4]. Steepest Ascent Hill Climbing [5] was a significant development, which added more systematic examination of neighbors. Stochastic variations [6, 7] were developed in the 1980s, which overcame the drawback of becoming trapped in local optima.

The simplicity and effectiveness of the algorithm have seen it become universally embraced in a range of applications, from early computer planning programs to complex optimization problems. Its influence can be seen in the development of more sophisticated algorithms such as Simulated Annealing [8] and Tabu Search [9, 10], whose basic principles build on Hill Climbing but remove its weaknesses.

10.1.3   Relationship to Other Optimization Techniques

Hill Climbing is a family of algorithms for local search. Unlike exhaustive search algorithms, it seeks to improve current solutions and not explore the entire search space. Some of its significant relations are:

1. Gradient Descent [11, 12]: Hill Climbing can be regarded as a discrete equivalent of gradient descent, with direction for improvement determined through examination of neighboring states, not computation of gradients.
2. Simulated Annealing [8, 13]: This method extends Hill Climbing by occasionally accepting worse solutions to escape local optima. The acceptance probability is controlled by a temperature parameter T:

$$P\left(\text{accept}\right) = e^{\frac{f(s') - f(s)}{T}}.$$

3. Genetic Algorithms [14]: While Hill Climbing modifies a single solution, genetic algorithms maintain a population of solutions that evolve through selection and mutation.

10.1.4   Key Applications in AI

Hill Climbing finds practical applications across various AI domains (Fig. 10.1).
   Despite its simplicity, Hill Climbing remains relevant in modern AI systems, particularly when:

- Quick approximate solutions are needed
- The optimization landscape is relatively smooth
- Computational resources are limited
- The problem has clearly defined neighboring states

The algorithm's limitations, particularly its tendency to get stuck in local optima, have led to various enhancements and hybrid approaches that we will explore in subsequent sections.

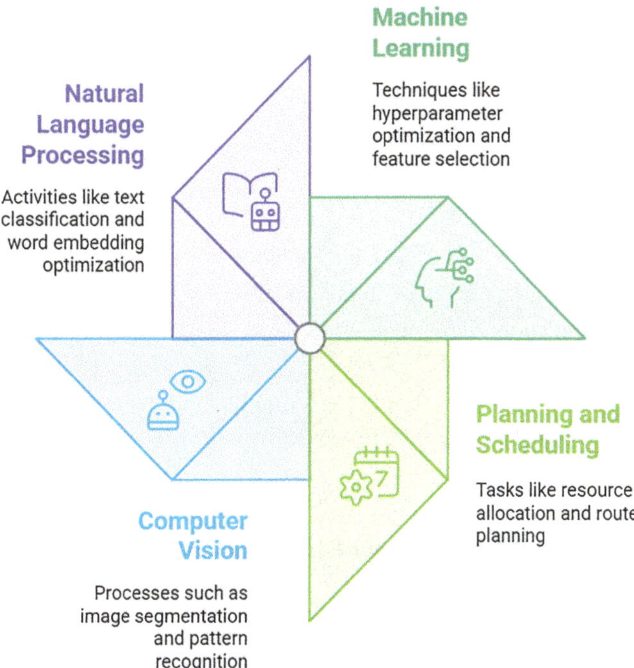

Fig. 10.1 Applications of Hill Climbing in AI

10.2   Theoretical Framework

The theoretical foundation of Hill Climbing rests on key concepts from optimization theory and search algorithms. This section explores these fundamental concepts to provide a rigorous understanding of how Hill Climbing operates.

10.2.1   Search Space and Objective Functions

A search space S represents all possible solutions to an optimization problem. Each point $s \in S$ corresponds to a potential solution. The objective function $f : S \to \mathrm{R}$ assigns a real value to each solution, quantifying its quality.

Formally, we express an optimization problem as:

$$\max_{s \in S} f(s).$$

For minimization problems, we can simply negate the objective function:

$$\min_{s \in S} f(s) = \max_{s \in S} -f(s).$$

Common examples of objective functions include:

1. Error minimization:

$$f(s) = -\sum_{i=1}^{n}(y_i - \hat{y}_i)^2.$$

2. Distance optimization:

$$f(s) = -\sum_{i=1}^{n} d(p_i, q_i).$$

3. Utility maximization:

$$f(s) = \sum_{i=1}^{n} u_i(s).$$

10.2.2   Local Optima vs Global Optima

A local optimum s_l is a solution that is better than all its neighbors:

$$f(s_l) \geq f(s), \forall s \in N(s_l),$$

where $N(s_l)$ represents the neighborhood of s_l.

A global optimum s_g is the best solution in the entire search space:

$$f(s_g) \geq f(s), \forall s \in S.$$

The distinction between local and global optima is crucial for understanding Hill Climbing's limitations. The algorithm may converge to a local optimum, missing the global optimum entirely.

10.2.3 *Neighborhood Functions*

The neighborhood function $N : S \rightarrow 2^S$ defines which solutions are accessible in a single step from the current solution. Common neighborhood definitions include:

1. For continuous spaces:

$$N(s) = s' : \| s' - s \| \leq \epsilon.$$

2. For discrete spaces:

$$N(s) = s' : d(s', s) = 1,$$

where d represents a suitable distance metric.

The choice of neighborhood function significantly impacts algorithm performance. A well-designed neighborhood function should:

- Maintain solution feasibility
- Allow sufficient exploration
- Be computationally efficient to evaluate

10.2.4 *Convergence Properties*

Hill Climbing's convergence depends on several factors:

1. Objective Function Properties:

   - Continuity
   - Smoothness
   - Number of local optima

2. Convergence Conditions:

$$s_{t+1} = s_t,$$

when

$$f(s') \leq f(s_t), \forall s' \in N(s_t).$$

3. Convergence Rate. For smooth functions near optima:

$$\| s_{t+1} - s_t \| \leq \alpha \| s_t - s_{t-1} \|,$$

where $\alpha < 1$ is the convergence rate.

Key convergence characteristics include:

1. Finite Termination. The algorithm always stops at a local optimum in finite steps for discrete spaces.
2. Quality Guarantee:

$$f(s_{final}) \geq f(s_0),$$

where s_{final} is the final solution and s_0 is the initial solution.
3. No Global Optimality. Hill Climbing cannot guarantee finding the global optimum unless the function is convex or unimodal.

10.3   Basic Hill Climbing Algorithm

10.3.1   Core Algorithm Structure

The basic Hill Climbing algorithm follows a simple iterative structure. At each step, it evaluates the current solution's neighbors and moves to the best improving neighbor. The process continues until no better neighbor exists.

The fundamental algorithm can be expressed in pseudocode:

```
function HILL-CLIMBING(problem) returns solution
   current ← INITIAL-STATE(problem)
   while true:
       neighbor ← BEST-NEIGHBOR(SUCCESSORS(problem, current))
       if VALUE(neighbor) ≤ VALUE(current):
           return current
       current ← neighbor
```

10.3.2   Implementation Considerations

Several key factors affect the algorithm's implementation:

1. Initial State Selection:

- Random initialization
- Heuristic-based initialization
- Multiple starting points

2. Neighbor Generation:

```python
def generate_neighbors(current_state, step_size):
    neighbors = []
    for dimension in range(len(current_state)):
        for step in [-step_size, step_size]:
            neighbor = current_state.copy()
            neighbor[dimension] += step
            neighbors.append(neighbor)
    return neighbors
```

3. Termination Conditions:

- No improvement found
- Maximum iterations reached
- Time limit exceeded
- Target value achieved

10.3.3 *Python Implementation Example*

Here's a complete implementation for continuous optimization problems:

```python
import numpy as np
def hill_climbing(objective_function,
                  initial_state,
                  step_size=0.1,
                  max_iterations=1000):
    """
    Parameters:
    -----------
    objective_function : callable
        Function to maximize
    initial_state : numpy.array
        Starting point
    step_size : float
        Size of steps in each dimension
    max_iterations : int
        Maximum number of iterations
```

```
Returns:
--------
tuple: (best_state, best_value)
"""
current_state = initial_state
current_value = objective_function(current_state)

for iteration in range(max_iterations):
    # Generate neighbors
    neighbors = []
    for i in range(len(current_state)):
        for step in [-step_size, step_size]:
            neighbor = current_state.copy()
            neighbor[i] += step
            neighbors.append(neighbor)

    # Evaluate neighbors
    best_neighbor = None
    best_neighbor_value = current_value

    for neighbor in neighbors:
        neighbor_value = objective_function(neighbor)
        if neighbor_value > best_neighbor_value:
            best_neighbor = neighbor
            best_neighbor_value = neighbor_value

    # Check if improvement found
    if best_neighbor_value <= current_value:
        break

    current_state = best_neighbor
    current_value = best_neighbor_value

return current_state, current_value
```

10.3.4   Step-by-Step Walkthrough

Let's analyze the algorithm's behavior on the Himmelblau function:

$$f(x,y) = -\left(x^2 + y - 11\right)^2 - \left(x + y^2 - 7\right)^2.$$

```python
def himmelblau(state):
    x, y = state
    return -(x**2 + y - 11)**2 - (x + y**2 - 7)**2
# Example usage
initial_state = np.array([1.0, 1.0])
result_state, result_value = hill_climbing(himmelblau,
                                            initial_state)
```

The algorithm proceeds through these steps:

1. Initialization: Start at point (1.0, 1.0)
2. Neighbor Generation: Create four neighbors at ±0.1 in each dimension
3. Evaluation: Calculate function value for each neighbor
4. Selection: Move to the best improving neighbor
5. Iteration: Continue until no improvement found

Monitoring progress reveals the optimization path:

```python
def monitor_hill_climbing(state, value, iteration):
    print(f"Iteration {iteration}:")
    print(f"State: {state}")
    print(f"Value: {value}\n")
```

This basic implementation demonstrates Hill Climbing's core principles while remaining simple enough to understand and modify for specific applications.

10.4   Variants of Hill Climbing

Several variants of Hill Climbing have been developed to address the limitations of the basic algorithm. Each variant offers different trade-offs between exploration and exploitation.

10.4.1   Steepest Ascent Hill Climbing

Steepest Ascent Hill Climbing examines all neighbors before making a move [5]. This variant ensures optimal local progress at each step.

Algorithm structure:

```python
def steepest_ascent_hill_climbing(problem):
    current = initial_state(problem)
    while True:
        neighbors = get_all_neighbors(current)
```

```
if not neighbors:
    return current
best = max(neighbors, key=evaluate)
if evaluate(best) <= evaluate(current):
    return current
current = best
```

The key difference from basic Hill Climbing lies in the evaluation of all neighbors:

$$s_{t+1} = \operatorname*{argmax}_{s' \in N(s_t)} f(s').$$

10.4.2   *Stochastic Hill Climbing*

This variant introduces randomness in neighbor selection [6, 7]. Instead of always choosing the best neighbor, it selects neighbors probabilistically based on their improvement.

The probability of selecting a neighbor is:

$$P(s') = \frac{e^{\beta f(s')}}{\displaystyle\sum_{s'' \in N(s)} e^{\beta f(s'')}},$$

where β controls the selection pressure.

Implementation example:

```python
def stochastic_hill_climbing(problem, beta=1.0):
    current = initial_state(problem)
    while True:
        neighbors = get_neighbors(current)
        if not neighbors:
            return current

        # Calculate selection probabilities
        values = [evaluate(n) for n in neighbors]
        probs = np.exp(beta * np.array(values))
        probs = probs / np.sum(probs)

        # Select neighbor probabilistically
        next_state = np.random.choice(neighbors, p=probs)
        if evaluate(next_state) <= evaluate(current):
            return current
        current = next_state
```

10.4.3   Random-Restart Hill Climbing

This variant addresses the local optima problem by running multiple searches from different starting points [15].
   Algorithm structure:

```python
def random_restart_hill_climbing(problem, n_restarts):
    best_solution = None
    best_value = float('-inf')

    for _ in range(n_restarts):
        solution = hill_climbing(problem,
                                 random_initial_state())
        value = evaluate(solution)

        if value > best_value:
            best_solution = solution
            best_value = value

    return best_solution
```

The probability of finding the global optimum increases with the number of restarts:

$$P\left(\text{finding global optimum}\right) = 1 - \left(1 - p\right)^{n},$$

where:

- p is the probability of reaching the global optimum from a single start
- n is the number of restarts

10.4.4   First-Choice Hill Climbing

This variant generates random neighbors until finding one that improves the current solution [16]. It is particularly effective for problems with many neighbors.
   Implementation:

```python
def first_choice_hill_climbing(problem, max_attempts=100):
    current = initial_state(problem)
    while True:
        found_better = False
```

```python
    for _ in range(max_attempts):
        neighbor = random_neighbor(current)
        if evaluate(neighbor) > evaluate(current):
            current = neighbor
            found_better = True
            break

    if not found_better:
        return current
```

10.4.5 Comparison of Hill Climbing Variants

Figure 10.2 illustrates the key characteristics and trade-offs of four main Hill Climbing variants.

Each variant offers distinct advantages and limitations that make it suitable for different optimization scenarios:

- Steepest Ascent prioritizes finding the best immediate improvement but requires more computational resources.
- Random-Restart focuses on finding better global solutions through multiple attempts.
- First-Choice offers computational efficiency with potentially suboptimal choices.
- Stochastic introduces randomness to escape local optima but risks missing good solutions.

These variants demonstrate how simple modifications to the basic Hill Climbing algorithm can significantly improve its performance in different scenarios.

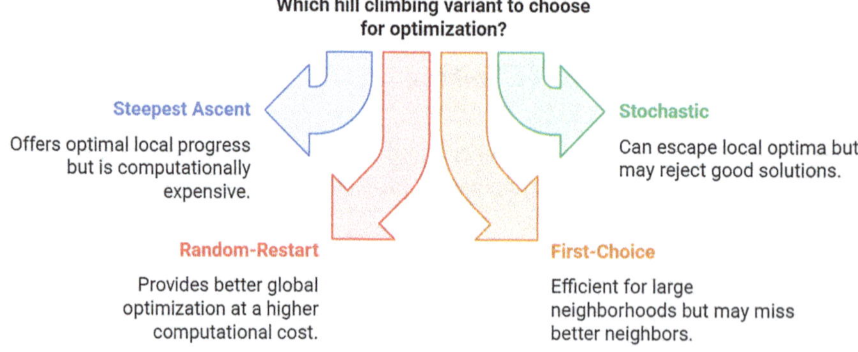

Fig. 10.2 Comparison of Hill Climbing variants: trade-offs and selection guide

10.5 Modern Applications and Advances

Recent studies revealed novel interesting new uses and enhancements of Hill Climbing algorithms:

- Urban Transportation Optimisation: Hill Climbing has proven to work remarkably well in solving big city transit routing problems. Latest studies illustrate that combining Hill Climbing with swarm intelligence algorithms can boost direct trip coverage by up to 11% in big city networks, and make it viable for real bus network planning [17].
- UAV Path Planning: Late Acceptance Hill Climbing (LAHC) has been a very successful technique for unmanned aerial vehicle path planning in complex urban environments. This variant suggests a memory-based strategy that significantly improves performance in terms of solution quality and convergence speed [18].
- Educational Scheduling: Steepest Ascent Hill Climbing, in a hyper-heuristic framework, worked optimally for university examination scheduling. Research attests to gains over traditional approaches, with lower penalty values for real-world applications [5].
- Evolutionary Algorithm Enhancement: There is a new algorithm, Balanced Hill Climbing Weight Algorithm with Diversity (BHWEAD), that couples Hill Climbing with evolutionary algorithms for an effective mix between exploration and exploitation. The hybrid algorithm works best in high-dimensional search spaces [19].
- Advanced Pivoting Methods: There have been new works that have developed expansion-based Hill Climbing, with maximimum expansion pivoting rules for choosing with a deeper level of information between improving neighbors. As a disadvantage, computational cost is heightened, but in certain instances, it heightens the level of local optima [20].

These advances confirm the long-term utility of Hill Climbing for modern-day optimization and its potential for expansion to new problem domains through innovative variants and hybrid approaches.

10.6 Practice and Exercises

10.6.1 Practice and Programming Exercises

All programming exercises for this chapter are available in the Google Colab notebook. You can access the complete notebook by scanning the QR code below or using this link: https://colab.research.google.com/drive/1XIQhSuSSTK1_r4D6Ha-4h5Mb_O4NWpUl

The notebook contains interactive Python code that you can run and modify to better understand Hill Climbing optimization concepts. Each exercise builds upon the previous ones, gradually introducing more complex aspects of optimization algorithms. Solutions to the exercises are available to instructors through the publisher's resources.

Exercise 1: Basic Hill Climbing Implementation (Notebook Cells 1–3)

Implement Hill Climbing for the Himmelblau function optimization problem. This exercise demonstrates the fundamental concepts through:

- Implementation of the basic algorithm
- Visualization of optimization paths
- Analysis of convergence behavior

Tasks:

- Generate synthetic Himmelblau function
- Implement basic Hill Climbing algorithm
- Visualize results in 2D and 3D
- Analyze optimization trajectories

Learning objectives:

- Understanding basic Hill Climbing implementation
- Implementing parameter updates
- Visualizing optimization results

Exercise 2: Variant Implementations (Notebook Cells 4–6)

Explore and implement different variants of Hill Climbing:

- Steepest Ascent Hill Climbing
- Stochastic Hill Climbing
- Random-Restart Hill Climbing

Tasks:

- Implement each variant
- Compare their performance
- Analyze convergence characteristics
- Visualize optimization paths

Exercise 3: Application Case Study (Notebook Cells 7–9)

Apply Hill Climbing to solve the Traveling Salesman Problem (TSP):

- TSP implementation
- Route optimization
- Performance comparison of variants

Tasks:

- Implement TSP problem structure
- Apply Hill Climbing variants
- Visualize and analyze results
- Compare performance metrics

Exercise 4: Performance Analysis (Notebook Cells 10–12)

Conduct comprehensive performance analysis:

- Convergence analysis
- Success rate evaluation
- Parameter sensitivity study
- Statistical comparison

Each exercise includes:

- Detailed code implementation
- Visualization components
- Analysis tools
- Performance metrics

The exercises provide hands-on experience with:

- Algorithm implementation
- Performance optimization
- Result visualization
- Comparative analysis

Expected Outputs: For each exercise, you should:

- Generate visualization plots
- Compare numerical results

- Analyze convergence behavior
- Document your findings

10.6.2   *Self-Assessment Questions*

Answer the following questions to test your understanding of Hill Climbing algorithms and their variants. Select the most appropriate answer for each question.

1. What is the primary principle behind Hill Climbing algorithms?

    (a) Random exploration of the search space
    (b) Moving to neighboring states with better values
    (c) Maintaining a population of solutions
    (d) Accepting worse solutions to escape local optima

2. In the context of Hill Climbing, what is a "plateau"?

    (a) The global maximum of the function
    (b) A region where neighboring states have equal values
    (c) The starting point of the algorithm
    (d) The steepest part of the search space

3. Why is Steepest Ascent Hill Climbing called "steepest"?

    (a) It only works on steep functions
    (b) It examines all neighbors before selecting the best one
    (c) It moves faster than other variants
    (d) It requires more computational resources

4. When might Random-Restart Hill Climbing be most useful?

    (a) When computation time is very limited
    (b) When the search space has many local optima
    (c) When memory is severely constrained
    (d) When the objective function is convex

5. What is the main advantage of Stochastic Hill Climbing over basic Hill Climbing?

    (a) It guarantees finding the global optimum
    (b) It has a chance to escape local optima
    (c) It requires less memory
    (d) It always converges faster

6. In the Traveling Salesman Problem context, what typically defines a "neighbor" solution?

    (a) A completely random new route
    (b) A route with two cities swapped
    (c) The shortest possible route
    (d) A route with all cities reversed

7. Which statement about Hill Climbing's convergence is correct?

   (a) It always finds the global optimum
   (b) It guarantees continuous improvement
   (c) It stops at the first local optimum found
   (d) It never gets stuck in local optima

8. What parameter is most crucial for basic Hill Climbing's performance?

   (a) Population size
   (b) Neighborhood function definition
   (c) Crossover rate
   (d) Mutation probability

Answers and Explanations:

1. (b) The fundamental principle of Hill Climbing is to iteratively move to better neighboring states.
2. (b) A plateau is a region where all neighboring states have equal values, making it difficult for the algorithm to determine the best direction.
3. (b) It examines all possible neighbors before making a move, ensuring the steepest improvement at each step.
4. (b) Random-Restart is particularly effective when dealing with multiple local optima, as it provides multiple chances to find better solutions.
5. (b) Stochastic Hill Climbing can potentially escape local optima by occasionally accepting moves to slightly worse states.
6. (b) In TSP, neighbors are typically defined by swapping two cities in the current route.
7. (c) Hill Climbing terminates when it reaches a local optimum, which may not be the global best solution.
8. (b) The definition of the neighborhood function critically determines which solutions are accessible in a single step.

These questions address key concepts and practical aspects of Hill Climbing algorithms. Understanding these fundamentals is essential for effectively implementing and applying Hill Climbing in real-world optimization problems.

10.7 Conclusion

Hill Climbing exemplifies how simple principles can yield powerful optimization tools. This chapter has examined the algorithm's theoretical foundations, practical implementations, and real-world applications.

We explored several algorithm variants. Steepest Ascent methodically evaluates all neighbors. Stochastic variants introduce beneficial randomness. Random-Restart approaches overcome local optima limitations. Each variant addresses specific weaknesses in the basic algorithm.

Several key factors influence Hill Climbing performance. The neighborhood function determines which solutions are accessible. The balance between exploration and exploitation affects convergence behavior. Problem representation significantly impacts optimization success. The match between algorithm variant and problem characteristics determines effectiveness.

These insights apply beyond Hill Climbing itself. They form building blocks for more sophisticated optimization techniques. Whether in neural network training or logistics optimization, these fundamental principles maintain their relevance.

References

1. Langley, P., Gennari, J.H., Iba, W.: Hill-Climbing Theories of Learning. In: Langley, P. (ed.) Proceedings of the Fourth International Workshop on MACHINE LEARNING. pp. 312–323. Morgan Kaufmann (1987). https://doi.org/10.1016/B978-0-934613-41-5.50035-0.
2. Tsai, C.-W., Chiang, M.-C.: Chapter Three—Traditional methods. In: Tsai, C.-W. and Chiang, M.-C. (eds.) Handbook of Metaheuristic Algorithms. pp. 29–69. Academic Press (2023). https://doi.org/10.1016/B978-0-44-319108-4.00016-2.
3. Storey, C.: Applications of a hill climbing method of optimization. Chemical Engineering Science. 17, 45–52 (1962). https://doi.org/10.1016/0009-2509(62)80005-0.
4. Rosenbrock, H.H., Storey, C.: CHAPTER 4—OPTIMIZING I—HILL-CLIMBING METHODS. In: Rosenbrock, H.H. and Storey, C. (eds.) Computational Techniques for Chemical Engineers. pp. 48–97. Pergamon (1966). https://doi.org/10.1016/B978-0-08-010889-6.50009-4.
5. Muklason, A., Pratama, E.J., Premananda, I.G.A.: Generic University Examination Timetabling System with Steepest-Ascent Hill Climbing Hyper-heuristic Algorithm. Procedia Computer Science. 234, 584–591 (2024). https://doi.org/10.1016/j.procs.2024.03.043.
6. Jacso, A., Lado, Z., Phanden, R.K., Sikarwar, B.S., Singh, R.K.: Bézier curve-based trochoidal tool path optimization using stochastic hill climbing algorithm. Materials Today: Proceedings. 78, 633–639 (2023). https://doi.org/10.1016/j.matpr.2022.12.056.
7. Mondal, B., Dasgupta, K., Dutta, P.: Load Balancing in Cloud Computing using Stochastic Hill Climbing-A Soft Computing Approach. Procedia Technology. 4, 783–789 (2012). https://doi.org/10.1016/j.protcy.2012.05.128.
8. Delahaye, D., Chaimatanan, S., Mongeau, M.: Simulated annealing: From basics to applications. Springer (2019). https://doi.org/10.1007/978-3-319-91086-4_1.
9. Tsai, C.-W., Chiang, M.-C.: Chapter Six—Tabu search. In: Tsai, C.-W. and Chiang, M.-C. (eds.) Handbook of Metaheuristic Algorithms. pp. 95–109. Academic Press (2023). https://doi.org/10.1016/B978-0-44-319108-4.00019-8.
10. Brandão, J.: A tabu search algorithm for the open vehicle routing problem. European Journal of Operational Research. 157, 552–564 (2004). https://doi.org/10.1016/S0377-2217(03)00238-8.
11. Tran-Dinh, Q., van Dijk, M.: Chapter 1—Gradient descent-type methods: Background and simple unified convergence analysis. In: Nguyen, L.M., Hoang, T.N., and Chen, P.-Y. (eds.) Federated Learning. pp. 3–28. Academic Press (2024). https://doi.org/10.1016/B978-0-44-319037-7.00008-9.
12. Zakwan, M.: Chapter 14—Gradient-based optimization. In: Eslamian, S. and Eslamian, F. (eds.) Handbook of Hydroinformatics. pp. 243–251. Elsevier (2023). https://doi.org/10.1016/B978-0-12-821285-1.00013-0.
13. Yang, X.-S.: Chapter 5—Simulated Annealing. In: Yang, X.-S. (ed.) Nature-Inspired Optimization Algorithms (Second Edition). pp. 83–90. Academic Press (2021). https://doi.org/10.1016/B978-0-12-821986-7.00012-3.

14. Tsai, C.-W., Chiang, M.-C.: Chapter Seven—Genetic algorithm. In: Tsai, C.-W. and Chiang, M.-C. (eds.) Handbook of Metaheuristic Algorithms. pp. 111–138. Academic Press (2023). https://doi.org/10.1016/B978-0-44-319108-4.00020-4.
15. Kato, E.R.R., Aranha, G.D. de A., Tsunaki, R.H.: A new approach to solve the flexible job shop problem based on a hybrid particle swarm optimization and Random-Restart Hill Climbing. Computers & Industrial Engineering. 125, 178–189 (2018). https://doi.org/10.1016/j.cie.2018.08.022.
16. Kratchanov, K., Golemanova, E., Golemanov, T., Ercan, T.: Non-procedural Implementation of Local Heuristic Search in Control Network Programming. (2010).
17. Zervas, A., Iliopoulou, C., Tassopoulos, I., Beligiannis, G.: Solving large-scale instances of the urban transit routing problem with a parallel artificial bee colony-hill climbing optimization algorithm. Applied Soft Computing. 167, 112335 (2024). https://doi.org/10.1016/j.asoc.2024.112335.
18. Deilam Salehi, E., Fazli, M.: Late acceptance hill climbing based algorithm for Unmanned Aerial Vehicles (UAV) path planning problem. Applied Soft Computing. 170, 112651 (2025). https://doi.org/10.1016/j.asoc.2024.112651.
19. Rodríguez-Esparza, E., Morales-Castañeda, B., Casas-Ordaz, A., Oliva, D., Navarro, M.A., Valdivia, A., Houssein, E.H.: Handling the balance of operators in evolutionary algorithms through a weighted Hill Climbing approach. Knowledge-Based Systems. 294, 111784 (2024). https://doi.org/10.1016/j.knosys.2024.111784.
20. Tari, S., Basseur, M., Goëffon, A.: Expansion-based Hill-climbing. Information Sciences. 649, 119635 (2023). https://doi.org/10.1016/j.ins.2023.119635.

Chapter 11
Simulated Annealing

Abstract This chapter explores Simulated Annealing (SA), a metaheuristic optimization technique inspired by metallurgical annealing. The text begins with the physical analogy of metal cooling to establish intuitive understanding. Mathematical foundations follow, covering temperature scheduling and acceptance criteria with clear explanations. Practical implementation is emphasized through Python code examples for both basic and advanced SA variants. The chapter examines applications ranging from the classic Traveling Salesman Problem to modern uses in VLSI design and portfolio optimization. Advanced topics include parallel implementations and adaptive cooling schedules. Each concept connects theory with practice through concrete examples and visualization techniques. The material provides both conceptual foundations and practical tools for applying SA to complex optimization problems.

11.1 Foundations of Simulated Annealing

Simulated Annealing (SA) is a powerful metaheuristic search algorithm that is inspired by the thermal metal annealing process in metallurgy [2, 3]. In this section, an introductory background and theoretical underpinnings for SA is discussed, offering a sound basis for an appreciation of its application in overcoming complex optimization problems.

11.1.1 Context and Motivation

Real-world optimization problems often involve finding solutions that minimize (or maximize) an objective function [4]. We can express this mathematically as finding a solution that achieves:

$$\min_{x \in S} f(x).$$

O. Kuznetsov, *Intelligent Systems: From Theory to Applications*, Cognitive
Technologies, https://doi.org/10.1007/978-3-032-00044-6_11

In this expression, $f : S \to R$ is the objective function to be optimized, and S is the solution space. x is a variable for a potential solution in our search space.

Traditional optimization methods face several significant challenges [4]. They often struggle with large search spaces. They frequently get trapped in local optima. Many real problems present non-convex objective functions. The solution space may be discrete or discontinuous. SA addresses these challenges through a probabilistic approach that permits temporary deterioration in solution quality, allowing escape from local optima.

11.1.2 *Physical Analogy*

The algorithm owes its name and its motivation to the metal annealing process in metallurgy [5, 6]. In metal annealing, a metal is first melted at a high temperature level. Temperature is then cooled down in a slow and controlled manner. In slow and controlled cooling, atoms have a chance to settle in a minimum-energy configuration state [7].

This physical analogue is a powerful metaphor for optimization. Energy level in the physical system corresponds to objective function value in our problem of optimization. Atomistic configuration corresponds to a feasible solution. Temperature corresponds to a control variable. Minimum energy level corresponds to our best solution.

11.1.3 *Core Principles*

The SA algorithm operates under three general principles [1, 2].

First, state transition is executed through the algorithm. In any move, it generates a neighboring solution through a small, random perturbation of its current state. For instance, in a Traveling Salesman Problem, it could swap two cities in its current path.

Second, the algorithm employs an acceptance criterion in terms of the Metropolis rule. In terms of such a criterion, an acceptance probability for a new solution

$$P(\text{accept}) = \begin{cases} 1 & \text{if } \Delta E \leq 0; \\ \exp(-\Delta E / T) & \text{if } \Delta E > 0. \end{cases}$$

Here, ΔE is objective value change between new and current solutions:

$$\Delta E = f\left(x_{\text{new}}\right) - f\left(x_{\text{current}}\right).$$

T is current temperature. Solutions with improvement in objective ($\Delta E \leq 0$) will have acceptance in an automatic manner. Poorer solutions can have acceptance, but with a probability that will decrease with a drop in temperature.

Third, the algorithm follows a temperature schedule. T_0 must be such that most of the transitions become feasible. Temperature decreases monotonically according to a cooling schedule. Simple scheme involves geometric cooling: $T_{k+1} = \alpha T_k$ with $0 < \alpha < 1$.

11.1.4 Mathematical Model

We can model behavior of the algorithm in terms of a Markov chain [1]. All feasible solutions S make up the state space. Transition probabilities depend on acceptance and mechanism of generation. For a constant temperature, in a stationary distribution, one obtains a Boltzmann distribution:

$$P(x) \propto \exp\left(-f(x)/T\right).$$

This mathematical underpinnings portrays several significant aspects of SA. Under proper cooling, it can arrive at global optima. It can tackle complex, multimodal objective functions with ease.

11.1.5 Comparison with Other Methods

SA offers several key benefits over conventional optimization techniques [1, 4]. In contrast to hill climbing, SA can escape a local optimum with its probabilistic acceptance criterion. It involves a moderate level of parameter tuning, balancing between the ease of use of random search and the high level of complexity of population-based techniques.

The algorithm provides probabilistic convergence guarantees under appropriate cooling schedules. It maintains low memory requirements, using only the current solution state. Perhaps most importantly, SA achieves a natural balance between exploration of the search space and exploitation of promising regions.

This theoretical foundation sets the stage for understanding both the convergence properties and practical implementation considerations. The subsequent sections will build upon these fundamentals to explore algorithmic variants, implementation strategies, and real-world applications.

11.2 Algorithm Design and Implementation: Central Concepts

The effective application of Simulated Annealing entails careful consideration of several key factors. All of them have a strong bearing on performance and efficiency in terms of discovering best-fit solutions. Let's cover them in a logical sequence.

11.2.1   State Representation

State representation forms the foundation of any SA implementation. A state represents a candidate solution in the solution space S. The choice of representation significantly impacts the algorithm's efficiency and effectiveness.

The representation must include all information pertinent to a solution but not become computationally infeasible. For the Traveling Salesman Problem, a permutation of cities is a natural one. In continuous optimization, state can be a real vector of values. For a problem in electronic circuits, state can represent positions and connectivity of parts.

11.2.2   Neighborhood Function

The neighborhood function $N(s)$ characterizes in what manner new candidate solutions can be produced out of a current state. Two important requirements have to be satisfied for $N(s)$. First, it should be possible to reach any solution in the search space through a sequence of neighborhood moves. Second, the moves should be relatively small to allow gradual exploration of the solution space.

For example, in the TSP, common neighborhood functions include:

- $N_{swap}(s)$: Exchange positions of two randomly chosen cities
- $N_{reverse}(s)$: Reverse the order of cities between two random positions

The choice of neighborhood function affects both the search trajectory and the algorithm's convergence properties.

11.2.3   Temperature Schedule

The temperature schedule controls the balance between exploration and exploitation throughout the search process. We typically define it through three components:

1. Initial Temperature (T_0): T_0 should be high enough to allow acceptance of most transitions initially. A common approach sets T_0 such that the initial acceptance ratio is approximately 0.8.
2. Cooling Function: The most widely used cooling function follows a geometric schedule:

$$T_{k+1} = \alpha T_k,$$

where α typically ranges from 0.9 to 0.99.

Alternative schedules include:

Linear cooling:

$$T_k = T_0 - \beta k.$$

Logarithmic cooling:

$$T_k = \frac{c}{\log(k+1)}.$$

3. Length of Markov Chain: At each temperature, we perform multiple transitions to approach equilibrium. The length often depends on problem size and complexity.

11.2.4   Acceptance Criterion

The Metropolis acceptance criterion determines whether to accept or reject a new solution [8, 9]. For a minimization problem, given current state s and candidate state s', the acceptance probability is:

$$P(s \to s') = \begin{cases} 1 & \text{if } f(s') \leq f(s); \\ \exp\left(\dfrac{f(s)-f(s')}{T}\right) & \text{otherwise.} \end{cases}$$

This criterion allows:

- Automatic acceptance of improving moves
- Probabilistic acceptance of deteriorating moves
- Decreasing probability of accepting worse solutions as temperature decreases

11.2.5   Termination Conditions

Several termination conditions can be employed, often in combination:

- Maximum Iterations: Stop after a predetermined number of iterations or temperature reductions.
- Temperature Threshold: Terminate when temperature falls below a specified value T_{min}.
- Solution Quality: Stop when a solution of acceptable quality is found or when no improvement occurs for a specified number of iterations.

where $d(x_i, x_j)$ is a distance between city i and city j. Neighborhood function will swap two arbitrary cities or reverse a portion of a route sequence.

VLSI Design

VLSI circuit design is yet another important application field [10, 11]. In this case, Simulated Annealing is utilized to minimize overall wire length and signal delay through optimizing chip component placement. In most cases, the objective function consists of several terms:

$$f(x) = \alpha L(x) + \beta C(x) + \gamma P(x),$$

where $L(x)$ is overall wire length, $C(x)$ is a measurement of circuit congestion, and $P(x)$ considers distribution of power. All three competing objectives are weighted with coefficients α, β, and γ.

Portfolio Optimization

Portfolio optimization employs Simulated Annealing for balancing between desired return and risk [12, 13]. Most often, its objective function conforms to the Markowitz model:

$$f(x) = -\left(\mu^T x - \lambda x^T \Sigma x\right),$$

where μ represents expected returns, Σ denotes the covariance matrix, and λ controls the risk-return trade-off. The algorithm must respect constraints like total investment and sector exposure limits.

Machine Learning

Machine learning involves training neural networks and feature selection. In feature selection, state is a feature selection vector with binary values representing desired features. The objective function aims for a prediction accuracy versus feature count tradeoff:

$$f(x) = Error(x) + \alpha \sum_{i=1}^{n} x_i,$$

where $Error(x)$ is an estimation of prediction error and α punishes the feature extraction count.

11.3.2   Advanced Techniques

Parallel Implementations

Parallel implementations significantly accelerate Simulated Annealing [14]. Multiple independent runs can explore different regions of the search space. One effective approach uses multiple Markov chains with periodic solution sharing:

$$x_i^{new} = \begin{cases} x_i^{local} & \text{with probability } p; \\ x_{best}^{global} & \text{with probability } 1-p. \end{cases}$$

Hybrid Approaches

Hybrid approaches combine Simulated Annealing with other optimization methods [12, 14]. A common strategy integrates local search to improve promising solutions:

1. Run Simulated Annealing to explore broadly
2. Apply local search to promising solutions
3. Update the current solution if improvement found
4. Continue with modified temperature schedule

Adaptive Temperature Scheduling

Adaptive temperature scheduling varies temperature according to search improvement [15, 16]. One successful technique varies the cooling schedule with regard to improvement in a solution:

$$T_{k+1} = \begin{cases} \alpha_{fast} T_k & \text{if improving;} \\ \alpha_{slow} T_k & \text{otherwise.} \end{cases}$$

Multi-Objective Optimization

Multi-objective optimization handles conflicting objectives [17, 18]. Pareto-based Simulated Annealing stores a set of non-dominated solutions. The acceptance probability is a function of relations of domination:

$$P(\text{accept}) = \begin{cases} 1 & \text{if } x_{new} \text{ dominates } x_{current}; \\ \exp\left(-\dfrac{d(x_{new}, x_{current})}{T}\right) & \text{otherwise,} \end{cases}$$

where $d(x_{new}, x_{current})$ measures the degree of non-domination.

These advanced algorithms make Simulated Annealing even more powerful in cases of complex optimizations. In most cases, however, problem-specific adaptations and careful parameter settings become a necessity. Technique selection is guided by specific requirements for an application, computational capabilities, and optimization objectives.

The field continues to develop with new adaptive approaches and blended methodologies. Most current work is focused in automatic parameter tuning and integration with machine learning algorithms. All of these advances make Simulated Annealing even more relevant for present deep learning, quantum computation, and sustainable energy systems.

11.4   Practice and Exercises

All programming exercises for this chapter are available in the Google Colab notebook. You can access the complete notebook by scanning the QR code below or using this link: https://colab.research.google.com/drive/1nA-PQl1vgWaEb6Gk25hbqj CU6Dp_L17b

The notebook contains interactive Python code that you can run and modify to better understand Simulated Annealing concepts. Each exercise builds upon the previous ones, gradually introducing more complex aspects of optimization algorithms. Solutions to the exercises are available to instructors through the publisher's resources.

11.4.1   Practice and Programming Exercises

Exercise 1: Basic Simulated Annealing Implementation

Implement basic Simulated Annealing algorithm with visualization capabilities. This exercise demonstrates the fundamental concepts of SA through a simple optimization problem.

Tasks:
- Implement the basic SA algorithm (see cells 1–2)
- Visualize the optimization process
- Experiment with different parameter settings
- Analyze convergence behavior

Learning Objectives:
- Understanding the basic SA algorithm implementation
- Grasping the role of temperature in optimization
- Mastering visualization of optimization progress

Exercise 2: TSP Solution with Simulated Annealing

Apply SA to solve the Traveling Salesman Problem, a classic combinatorial optimization challenge.

Tasks:
- Implement TSP-specific components (cells 3–4)
- Generate test data and calculate distances
- Visualize routes and optimization progress
- Compare solutions with different parameter settings

Learning Objectives:
- Applying SA to a specific optimization problem
- Understanding problem-specific adaptations
- Analyzing solution quality and convergence

Exercise 3: Parameter Analysis

Explore how different parameters affect SA performance and solution quality.

Tasks:
- Analyze impact of initial temperature (cell 5)
- Study cooling schedule effects
- Evaluate convergence behavior
- Compare different parameter combinations

Learning Objectives:
- Understanding parameter sensitivity
- Mastering parameter tuning strategies
- Analyzing optimization trade-offs

Exercise 4: Advanced SA Variants

Implement and compare advanced versions of the Simulated Annealing algorithm.

Tasks:
- Implement parallel SA (cell 6)
- Add adaptive cooling mechanisms
- Compare performance with basic SA
- Visualize comparative results

Learning Objectives:
- Understanding advanced SA techniques
- Implementing parallel optimization
- Evaluating algorithm improvements

11.4.2   Self-Assessment Questions

Please answer the following questions to evaluate your understanding of Simulated Annealing. Select the answer you think is correct for each question.

1. What physical process inspired the Simulated Annealing algorithm?

    (a) Formation of crystals
    (b) Metal annealing process
    (c) Nuclear fusion
    (d) Water evaporation

2. In Simulated Annealing, what does temperature control?

    (a) The speed of the algorithm
    (b) The size of the search space
    (c) The probability of accepting worse solutions
    (d) The number of iterations

3. How does the algorithm handle worse solutions during the search process?

    (a) Always rejects them
    (b) Always accepts them
    (c) Accepts them with a probability based on temperature
    (d) Only accepts them at high temperatures

4. What happens to the acceptance probability of worse solutions as temperature decreases?

    (a) Increases linearly
    (b) Remains constant
    (c) Decreases exponentially
    (d) Increases exponentially

5. Which of the following is NOT a common cooling schedule in Simulated Annealing?

    (a) Linear cooling
    (b) Geometric cooling

   (c) Logarithmic cooling

   (d) Exponential heating

6. What role does the initial temperature play in Simulated Annealing?

   (a) Determines the final solution quality

   (b) Controls the total number of iterations

   (c) Sets the initial exploration capability

   (d) Defines the problem constraints

7. When implementing Simulated Annealing for the TSP, which operation is typically used to generate neighbor solutions?

   (a) Adding new cities

   (b) Swapping pairs of cities

   (c) Removing cities

   (d) Changing city coordinates

8. What is the main advantage of parallel Simulated Annealing?

   (a) Lower memory usage

   (b) Simpler implementation

   (c) Broader exploration of solution space

   (d) Guaranteed optimal solution

9. What is the typical stopping criterion for Simulated Annealing?

   (a) When the temperature reaches zero

   (b) After a fixed number of iterations

   (c) When a perfect solution is found

   (d) When memory is exhausted

10. Which statement about Simulated Annealing is correct?

   (a) It always finds the global optimum

   (b) It works only on discrete problems

   (c) It balances exploration and exploitation

   (d) It requires no parameter tuning

Correct answers:

1. (b)
2. (c)
3. (c)
4. (c)
5. (d)
6. (c)
7. (b)
8. (c)
9. (b)
10. (c)

11.5 Conclusion

Simulated Annealing provides a robust framework for solving complex optimization problems across diverse domains. Its strength lies in balancing exploration and exploitation through temperature-controlled probabilistic acceptance of solutions. This balance allows SA to escape local optima while gradually converging toward global optimal solutions.

The key contributions of SA to optimization include:

- Ability to handle complex, non-convex optimization landscapes
- Relatively simple implementation with modest parameter tuning requirements
- Adaptability to various problem domains through problem-specific neighborhood functions
- Natural extension to parallel and adaptive variants for enhanced performance

Current research continues to extend SA's capabilities through:

- Integration with machine learning for automatic parameter tuning
- Applications in quantum computing optimization
- Hybrid approaches combining SA with other metaheuristics
- Development of more efficient cooling schedules for specific problem classes

The principles and techniques presented in this chapter provide a foundation for applying SA to practical optimization challenges across engineering, finance, and artificial intelligence domains.

References

1. Delahaye, D., Chaimatanan, S., Mongeau, M.: Simulated annealing: From basics to applications. Springer (2019). https://doi.org/10.1007/978-3-319-91086-4_1.
2. Kirkpatrick, S., Gelatt, C.D., Vecchi, M.P.: Optimization by simulated annealing. Science. 220, 671–680 (1983). https://doi.org/10.1126/science.220.4598.671.
3. Kirkpatrick, S.: Optimization by simulated annealing: Quantitative studies. J Stat Phys. 34, 975–986 (1984). https://doi.org/10.1007/BF01009452.
4. Eremia, M., Liu, C.-C., Edris, A.-A.: Heuristic Optimization Techniques. In: Advanced Solutions in Power Systems: HVDC, FACTS, and Artificial Intelligence. pp. 931–984. IEEE (2016). https://doi.org/10.1002/9781119175391.ch21.
5. Tsai, C.-W., Chiang, M.-C.: Chapter Five—Simulated annealing. In: Tsai, C.-W. and Chiang, M.-C. (eds.) Handbook of Metaheuristic Algorithms. pp. 79–93. Academic Press (2023). https://doi.org/10.1016/B978-0-44-319108-4.00018-6.
6. Yang, X.-S.: Chapter 5—Simulated Annealing. In: Yang, X.-S. (ed.) Nature-Inspired Optimization Algorithms (Second Edition). pp. 83–90. Academic Press (2021). https://doi.org/10.1016/B978-0-12-821986-7.00012-3.
7. Landau, L.D., Lifshitz, E.M.: Statistical Physics: Volume 5. Elsevier (2013).
8. Metropolis, N., Rosenbluth, A.W., Rosenbluth, M.N., Teller, A.H., Teller, E.: Equation of State Calculations by Fast Computing Machines. J. Chem. Phys. 21, 1087–1092 (1953). https://doi.org/10.1063/1.1699114.

9. Klenke, A.: Wahrscheinlichkeitstheorie. Springer Spektrum, Berlin; [Heidelberg] (2020). https://doi.org/10.1007/978-3-662-62089-2.

10. Kumar, S.B.V., Rao, P.V., Sharath, H.A., Sachin, B.M., Ravi, U.S., Monica, B.V.: Review on VLSI design using optimization and self-adaptive particle swarm optimization. Journal of King Saud University—Computer and Information Sciences. 32, 1095–1107 (2020). https://doi.org/10.1016/j.jksuci.2018.01.001.

11. Sait, S.M., Khan, J.A.: Simulated evolution for timing and low power VLSI standard cell placement. Engineering Applications of Artificial Intelligence. 16, 407–423 (2003). https://doi.org/10.1016/j.engappai.2003.08.004.

12. Gunjan, A., Bhattacharyya, S.: Chapter 7—Portfolio optimization using simulated annealing and quantum-inspired simulated annealing: A comparative study. In: Bhattacharyya, S., Köppen, M., De, D., and Panigrahi, B.K. (eds.) Recent Trends in Swarm Intelligence Enabled Research for Engineering Applications. pp. 213–243. Academic Press (2024). https://doi.org/10.1016/B978-0-443-15533-8.00014-X.

13. Som, A., Kayal, P.: A multicountry comparison of cryptocurrency *vs* gold: Portfolio optimization through generalized simulated annealing. Blockchain: Research and Applications. 3, 100075 (2022). https://doi.org/10.1016/j.bcra.2022.100075.

14. Hernández, J., Minetti, G., Salto, C., Carnero, M., Sánchez, M.: Parallel Simulated Annealing approach for optimal process plants instrumentation. In: Montastruc, L. and Negny, S. (eds.) Computer Aided Chemical Engineering. pp. 1285–1290. Elsevier (2022). https://doi.org/10.1016/B978-0-323-95879-0.50215-0.

15. Cheng, L., Tang, Q., Zhang, L.: Mathematical model and adaptive simulated annealing algorithm for mixed-model assembly job-shop scheduling with lot streaming. Journal of Manufacturing Systems. 70, 484–500 (2023). https://doi.org/10.1016/j.jmsy.2023.08.008.

16. Moreno, F., Forcael, E., Orozco, F., Baesler, F., Agdas, D.: Fixed start method for repetitive project scheduling with simulated annealing. Heliyon. 11, e41741 (2025). https://doi.org/10.1016/j.heliyon.2025.e41741.

17. Wang, F., Chun, W., Wu, W.: Application of simulated annealing algorithm in multi-objective allocation optimization of urban water resources. Desalination and Water Treatment. 314, 304–313 (2023). https://doi.org/10.5004/dwt.2023.30032.

18. Motaghedi-Larijani, A.: Solving the number of cross-dock open doors optimization problem by combination of NSGA-II and multi-objective simulated annealing. Applied Soft Computing. 128, 109448 (2022). https://doi.org/10.1016/j.asoc.2022.109448.

Chapter 12
Gradient-Based Optimization

Abstract This chapter examines gradient-based optimization methods, essential tools in modern machine learning and artificial intelligence. We extend previous optimization approaches to continuous spaces, showing how derivatives guide the search process toward optimal solutions. The material progresses from fundamental gradient descent to sophisticated adaptive techniques, with emphasis on neural network training applications. Key concepts include learning rate dynamics, momentum-based acceleration, and specialized algorithms like Adam and RMSprop. Each method is presented with mathematical foundations, implementation details, and convergence analysis. Practical Python implementations demonstrate algorithm behavior on both classical optimization problems and contemporary machine learning tasks. The chapter concludes with emerging research directions and advanced techniques for large-scale systems. Through this structured approach, readers gain both theoretical understanding and practical implementation skills for applying gradient-based methods to complex optimization challenges.

12.1 Foundations of Gradient-Based Optimization

Gradient-based optimization stands out as a cornerstone of modern artificial intelligence and machine learning [1, 2]. In contrast to previously studied methods, which are either based on discrete search spaces such as in the case of A* or do not leverage gradient information during the search such as in Hill Climbing and Simulated Annealing, gradient-based methods are meant for operating on continuous spaces using derivative information for driving the process.

12.1.1 Context and Mathematical Foundations

In its most basic form, we can express an optimization problem as [2, 3]:

247

O. Kuznetsov, *Intelligent Systems: From Theory to Applications*, Cognitive
Technologies, https://doi.org/10.1007/978-3-032-00044-6_12

11.2.6   Implementation Framework

The following pseudocode encapsulates these components:

```
Initialize(s0, T0)
s = s0   // Current solution
sbest = s0   // Best solution found
T = T0   // Initial temperature
while (not terminated) do
   for i = 1 to MarkovChainLength do
       s' = GenerateNeighbor(s)
       ΔE = f(s') - f(s)
       if Random(0,1) < exp(-ΔE/T) then
           s = s'
           if f(s) < f(sbest) then
               sbest = s
   T = UpdateTemperature(T)
return sbest
```

The integration of such modules creates a robust optimization algorithm. Individual modules can then be tuned for a problem structure with no loss in the basic principles of simulated annealing. In the following section, we will present practical techniques and settings for tuning.

11.3   Applications and Advanced Topics

This section covers real implementations and complex variants of Simulated Annealing and its real-life applications, taking its capabilities to new dimensions with new techniques.

11.3.1   Practical Applications

Traveling Salesman Problem

The Traveling Salesman Problem (TSP) is a classic application of Simulated Annealing [1, 3]. In TSP, one wishes to find the shortest path that covers each city and comes back to the starting city, visiting each city but once. The objective function is to obtain the overall distance:

$$f(x) = \sum_{i=1}^{n-1} d(x_i, x_{i+1}) + d(x_n, x_1),$$

$$\min_{x \in \mathbb{R}^n} f(x),$$

where $f: \mathbb{R}^n \to \mathbb{R}$ is the objective function, we want to minimize. This formulation appears throughout machine learning:

- In neural networks: minimizing the loss function;
- In regression: minimizing the mean squared error;
- In classification: minimizing cross-entropy loss.

The gradient of function f at point $x = (x_1, \ldots, x_n)$ is defined as the vector of partial derivatives:

$$\nabla f(x) = \left(\frac{\partial f}{\partial x_1}, \ldots, \frac{\partial f}{\partial x_n} \right).$$

12.1.2   Geometric Interpretation

The gradient vector has two crucial properties that make it valuable for optimization:

- Direction: The gradient points in the direction of steepest increase of the function
- Magnitude: The gradient's magnitude indicates how steep the function is at that point

Therefore, moving in the opposite direction of the gradient $(-\nabla f(x))$ leads us toward the function's minimum. This simple but powerful insight forms the basis of gradient-based optimization methods.

12.1.3   Relationship to Previous Optimization Methods

Let's compare gradient-based optimization with previously studied methods (Table 12.1).

Gradient-based optimization methods are particularly effective for continuous optimization problems, especially in high-dimensional settings, making them indispensable in machine learning. However, in combinatorial optimization tasks, methods such as A* or simulated annealing may be more appropriate due to their heuristic or probabilistic nature.

12.1.4   Role in Machine Learning and AI

Gradient-based optimization methods are especially important when solving complex optimization problems (see Fig. 12.1).

Table 12.1 Comparison of optimization methods

Method	Search space	Uses gradient information	Resistance to local optima	Efficiency in high dimensions	Remarks
A* Algorithm	Discrete	No	Guarantees optimality (with an admissible heuristic)	Not applicable	A heuristic-based shortest path search algorithm
Hill Climbing	Discrete and continuous	No	Susceptible to local optima	Often inefficient	A simple greedy search algorithm
Simulated Annealing	Discrete and continuous	No	Can escape local optima via probabilistic moves	Poor scalability in continuous optimization	Utilizes probabilistic transitions to enhance solution quality
Gradient-Based Methods	Continuous	Yes	Depends on function properties	Highly efficient	Fundamental in modern deep learning optimization

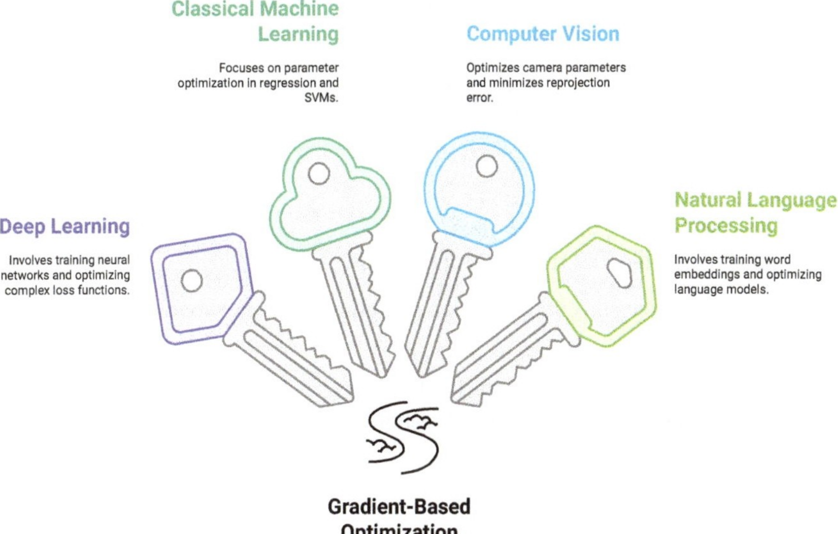

Fig. 12.1 Harnessing gradient optimization across AI domains

The widespread success of gradient-based methods in these areas stems from their ability to efficiently navigate high-dimensional spaces by following the local geometry of the objective function. However, they also face challenges such as:

- Getting stuck in local minima;
- Dealing with saddle points;

- Managing the learning rate;
- Handling non-convex optimization landscapes.

These challenges have led to the development of more sophisticated variants, which we will explore in subsequent sections.

12.2   The Gradient Descent Algorithm

Gradient Descent (GD) is the foundational algorithm for gradient-based optimization. Its principle is remarkably simple: iteratively move in the direction opposite to the gradient to find the minimum of a function. This section explores both the theoretical foundations and practical implementations of this powerful algorithm.

12.2.1   Basic Algorithm

The core gradient descent algorithm can be expressed in a single update equation [1, 2]:

$$x_{t+1} = x_t - \eta \nabla f\left(x_t\right),$$

where:

- x_t is the current point at iteration t;
- $\eta > 0$ is the learning rate (step size);
- $\nabla f(x_t)$ is the gradient of function f at point x_t.

The algorithm repeats this update until a stopping criterion is met, such as:

- Maximum number of iterations reached;
- Gradient magnitude below a threshold;
- Change in function value below a threshold.

12.2.2   The Learning Rate

The learning rate η is arguably the most critical hyperparameter in gradient descent. It determines how large a step we take in the direction of the negative gradient (Fig. 12.2).

The choice of learning rate (η) significantly influences the optimization process (Fig. 12.2). A learning rate that is too large can cause instability and divergence, while a too-small value leads to slow convergence and inefficiency. Selecting an optimal learning rate is crucial for balancing speed and stability, often requiring empirical tuning or adaptive techniques.

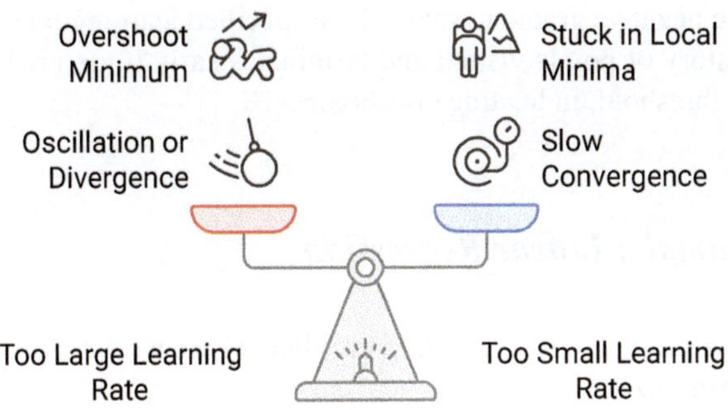

Fig. 12.2 Impact of learning rate on optimization performance

12.2.3 Implementation in Python

Here's a basic implementation of gradient descent:

```python
def gradient_descent(f, grad_f, x0, learning_rate, n_iterations):
    """
    Parameters:
    f: objective function
    grad_f: gradient function
    x0: initial point
    learning_rate: step size
    n_iterations: maximum number of iterations
    """
    x = x0
    history = []
    for i in range(n_iterations):
        # Store current point
        history.append((x, f(x)))
        # Compute gradient
        grad = grad_f(x)
        # Update x
        x = x - learning_rate * grad
        # Optional: stopping criterion
        if np.linalg.norm(grad) < 1e-6:
            break
    return x, history
```

This function implements gradient descent, used to minimize a given function
$f(x)$. It starts from an initial point x_0 and iteratively updates x by moving in the

direction of the negative gradient, scaled by a specified learning rate. The algorithm records the history of points visited and terminates early if the gradient norm falls below a small threshold, indicating convergence.

12.2.4   Example: Linear Regression

Let's consider a simple linear regression problem where we want to find parameters w and b that minimize:

$$f(w,b) = \frac{1}{2n} \sum_{i=1}^{n} (wx_i + b - y_i)^2.$$

The gradients are:

$$\frac{\partial f}{\partial w} = \frac{1}{n} \sum_{i=1}^{n} (wx_i + b - y_i) x_i,$$

$$\frac{\partial f}{\partial b} = \frac{1}{n} \sum_{i=1}^{n} (wx_i + b - y_i).$$

Implementation example:

```python
def linear_regression_gradient_descent(X, y, learning_rate=0.1,
n_iterations=1000):
w = 0
b = 0
n = len(X)
for i in range(n_iterations):
    # Compute gradients
    y_pred = w * X + b
    error = y_pred - y
    grad_w = np.mean(error * X)
    grad_b = np.mean(error)
    # Update parameters
    w = w - learning_rate * grad_w
    b = b - learning_rate * grad_b
return w, b
```

It aims to find the best slope www and intercept bbb that minimize the error between predictions and actual values. The algorithm starts with $w = 0$ and $b = 0$, then iteratively updates them. It calculates the prediction error and adjusts www and bbb using the gradients. The learning rate controls the step size of updates. Over time, the parameters improve, leading to a better fit of the data.

Table 12.2 Factors affecting gradient descent convergence

Factor	Effect on convergence	Remarks
Function Properties	Convex: Converges to the global minimum	Guarantees optimal solution in convex cases
	Non-convex: May converge to a local minimum	Solution quality depends on the function landscape
	Smoothness and Lipschitz continuity influence the convergence rate	Well-conditioned functions improve stability
Learning Rate	Must meet specific conditions for convergence	Improper tuning can lead to divergence or slow learning
	Too large: May not converge	Causes instability and oscillations
	Too small: Converges very slowly	Results in inefficient optimization
Initial Point	Determines which local minimum is found	Important in non-convex functions
	Influences convergence speed	A poor choice can slow down learning

12.2.5   Convergence Properties

Gradient descent's convergence depends on several factors (Table 12.2).

Table 12.2 summarizes key factors influencing gradient descent convergence. Function properties determine whether the algorithm finds a global or local minimum. The learning rate must be carefully chosen to balance speed and stability. The initial point is particularly crucial in non-convex settings, as it affects both the final solution and convergence rate.

12.2.6   Common Challenges and Solutions

Gradient descent is a powerful optimization method, but its performance depends on several factors. Figure 12.3 summarizes common challenges and practical solutions to improve convergence and stability:

- Gradient descent works best when the learning rate is well-tuned. A learning rate that is too high can cause divergence, while a very low learning rate results in slow convergence. To address this, learning rate scheduling and adaptive methods help maintain an optimal step size throughout training.
- Local minima can be problematic in non-convex optimization, but multiple initializations, momentum-based techniques, and stochastic variations can help avoid poor solutions. Slow convergence is another issue, particularly in deep learning, where adaptive optimizers and proper feature scaling improve training efficiency.
- Finally, ill-conditioning, where small parameter changes lead to large fluctuations in the loss function, can significantly slow down convergence. Normalization,

Fig. 12.3 Common challenges in gradient descent and their solutions

preconditioning, and second-order methods help mitigate this problem by improving numerical stability.

By understanding and addressing these challenges, gradient descent can be made more robust and efficient, leading to better optimization results.

12.3   Advanced Gradient Descent Variants

While simple gradient descent is a good starting point for optimization, in practice and with more modern machine learning, one often has to go for more sophisticated methods. The section presents some advanced variants that address different shortcomings of the basic algorithm.

12.3.1   Stochastic Gradient Descent (SGD)

Stochastic Gradient Descent modifies the standard gradient descent by computing the gradient using a single randomly selected example at each iteration. For a loss function $L(\theta)$ with n training examples:

$$L(\theta) = \frac{1}{n}\sum_{i=1}^{n} L_i(\theta).$$

The SGD update rule becomes:

$$\theta_{t+1} = \theta_t - \eta \nabla L_i(\theta_t),$$

where i is randomly selected at each iteration. This approach offers several advantages:

- Computational efficiency for large datasets
- Ability to escape local minima through noise in gradient estimation
- Online learning capability for streaming data

However, SGD also introduces higher variance in the parameter updates, which can make convergence more erratic.

12.3.2   Mini-Batch Gradient Descent

Mini-batch gradient descent represents a compromise between batch and stochastic gradient descent. It updates parameters using a small random subset of training examples:

$$\theta_{t+1} = \theta_t - \eta \frac{1}{|B_t|}\sum_{i\in B_t} \nabla L_i(\theta_t),$$

where B_t is the mini-batch at iteration t. The batch size typically ranges from 32 to 512 examples, balancing between computational efficiency and update stability.

Python implementation example:

```python
def minibatch_gradient_descent(X, y, batch_size=32,
learning_rate=0.01):
n_samples = len(X)
theta = initialize_parameters()
for epoch in range(n_epochs):
    # Shuffle data
    indices = np.random.permutation(n_samples)
    # Process mini-batches
    for start_idx in range(0, n_samples, batch_size):
        batch_idx = indices[start_idx:start_idx + batch_size]
        X_batch = X[batch_idx]
        y_batch = y[batch_idx]
```

```
    # Compute gradient on mini-batch
    grad = compute_gradient(X_batch, y_batch, theta)

    # Update parameters
    theta = theta - learning_rate * grad

return theta
```

This function implements mini-batch gradient descent, an optimization technique that balances efficiency and stability by updating parameters using small random subsets of the data. It runs through numerous iterations in the shuffle across datasets that are supposed to make a better convergence; inside of these epochs, mini-batch processing with computations of gradients and model parameters' update happens, thus being the methods reducing the variance at a comparable computational efficiency with the stochastic gradient descent.

12.3.3   Momentum-Based Methods

Momentum methods address the high variance in SGD by accumulating a moving average of past gradients. The classical momentum update is:

$$v_{t+1} = \gamma v_t + \eta \nabla L(\theta_t),$$

$$\theta_{t+1} = \theta_t - v_{t+1},$$

where $\gamma \in [0, 1]$ is the momentum coefficient. This approach provides several benefits:

- Accelerated convergence in ravines
- Reduced oscillation in high-curvature directions
- Ability to overcome small local minima

A popular variant is Nesterov Accelerated Gradient (NAG):

$$v_{t+1} = \gamma v_t + \eta \nabla L(\theta_t - \gamma v_t),$$

$$\theta_{t+1} = \theta_t - v_{t+1}.$$

12.3.4   Understanding Algorithm Behavior

The choice of gradient descent method directly impacts training efficiency and convergence behavior (Fig. 12.4).

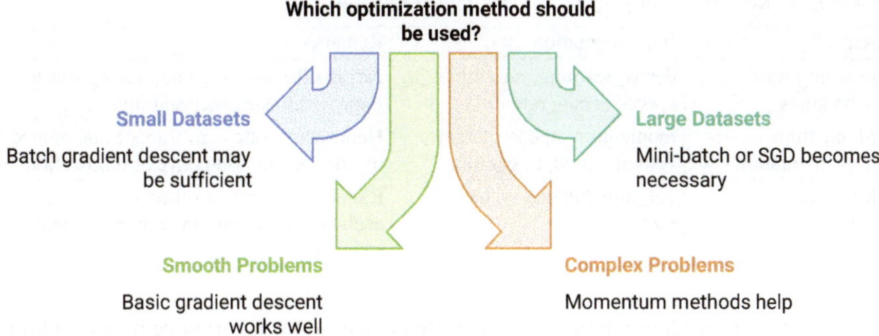

Fig. 12.4 Choosing the appropriate gradient descent variant

For small datasets, batch gradient descent works fine, since it provides stable updates. For large datasets, one has to resort to mini-batch gradient descent or stochastic gradient descent to save on memory and computational cost.

Another important factor is the nature of the optimization problem. While simple convex functions can be optimized effectively using basic gradient descent, complex non-convex landscapes require advanced techniques that efficiently handle saddle points and local minima, such as momentum-based optimizers.

Finally, the available hardware resources play a role. If the memory is limited, the choice of SGD is preferable since it requires less amount of storage, while a mini-batch gradient descent can take advantage of parallel computation on GPUs and thus significantly speed up the training. All these factors taken into consideration will allow the selection of the most appropriate optimization strategy for the problem in question.

12.3.5 Practical Implementation Considerations

Effective implementation of gradient descent requires careful handling of learning rate scheduling, momentum, and mini-batch sampling:

1. Learning Rate Scheduling: Learning rate scheduling is essential for stabilizing training. A high initial learning rate allows for rapid progress, while gradual decay prevents overshooting the optimum. The function `decay_learning_rate` implements a simple decay strategy by reducing the learning rate over time.

```python
def decay_learning_rate(initial_lr, epoch, decay_rate=0.1):
return initial_lr / (1 + decay_rate * epoch)
```

Table 12.3 Key aspects of gradient descent implementation

Aspect	Implementation	Remarks
Learning Rate Scheduling	decay_learning_rate(initial_lr, epoch, decay_rate=0.1)	Adjusts the learning rate over epochs to improve convergence stability
Momentum Implementation	momentum_update(params, velocity, grad, lr, gamma)	Helps accelerate convergence and reduce oscillations in non-convex optimization
Mini-batch Sampling	get_minibatch(X, y, batch_size)	Randomly selects a subset of data for each update, improving computational efficiency

2. Momentum Implementation: Momentum accelerates convergence by smoothing updates, reducing the impact of noisy gradients. The function `momentum_update` maintains a velocity term that accumulates past gradients, allowing faster movement in consistent directions while damping oscillations.

```
def momentum_update(params, velocity, grad, lr, gamma):
velocity = gamma * velocity + lr * grad
params = params - velocity
return params, velocity
```

3. Mini-batch Sampling: Mini-batch sampling is crucial for efficient training on large datasets. The function `get_minibatch` randomly selects a subset of data points, ensuring that each update step is computed efficiently without requiring full dataset evaluation.

```
def get_minibatch(X, y, batch_size):
indices = np.random.choice(len(X), batch_size, replace=False)
return X[indices], y[indices]
```

Table 12.3 summarizes these key aspects and their impact on optimization performance.

By carefully implementing these components, gradient descent can be optimized for faster, more stable training.

12.3.6   Performance Comparison

Different gradient descent variants exhibit distinct trade-offs in terms of convergence speed, memory usage, and computational efficiency. Table 12.4 summarizes these characteristics, helping to determine the most suitable approach based on the problem requirements.

The choice of gradient descent variant depends on the dataset size, available computational resources, and optimization landscape:

Table 12.4 Comparison of gradient descent variants

Criterion	Batch gradient descent (GD)	Stochastic gradient descent (SGD)	Mini-batch gradient descent	Remarks
Convergence Speed	Stable but slow	Fast initial progress, but noisy	Balanced speed and stability	SGD may oscillate, while mini-batch stabilizes learning
Memory Requirements	High (stores full dataset)	Low (processes one sample at a time)	Adjustable via batch size	Mini-batch allows control over memory usage
Computational Efficiency	High (full dataset per update)	Efficient per update but frequent updates	Efficient with parallelization	Mini-batch optimizes GPU usage for large datasets

- Batch gradient descent ensures smooth convergence but is computationally expensive and memory-intensive.
- SGD progresses rapidly by updating after each sample but suffers from instability due to noisy updates.
- Mini-batch gradient descent provides a compromise, reducing noise while maintaining efficiency, making it particularly suitable for large-scale machine learning tasks.

Proper selection of batch size and momentum further improves optimization performance.

12.4   Modern Optimization Algorithms

Modern deep learning is based on sophisticated algorithms of optimization, most of which constitute an extension of the basic method of gradient descent. The section covers the most successful modern optimizers, their mathematical grounds, and practical implementations.

12.4.1   AdaGrad: Adaptive Gradient Algorithm

AdaGrad adapts the learning rate for each parameter based on the history of gradient updates [4, 5]. This makes it particularly effective for sparse data and different scales of features.

The update rule for AdaGrad is:

$$G_{t,ii} = \sum_{\tau=1}^{t} \left(g_{\tau,i} \right)^2,$$

$$\theta_{t+1,i} = \theta_{t,i} - \frac{\eta}{\sqrt{G_{t,ii}} + \epsilon} g_{t,i},$$

where:

- $g_{t,i}$ is the gradient for parameter i at time t;
- $G_{t,ii}$ accumulates squared gradients;
- ϵ is a small constant to prevent division by zero.

Python implementation:

```python
class AdaGrad:
def __init__(self, params, learning_rate=0.01, epsilon=1e-8):
    self.params = params
    self.lr = learning_rate
    self.epsilon = epsilon
    self.G = {param: np.zeros_like(param) for param in params}
def step(self, grads):
    for param, grad in zip(self.params, grads):
        self.G[param] += grad ** 2
        param -= self.lr * grad / (np.sqrt(self.G[param] + self.
epsilon))
```

This class implements AdaGrad, that adjusts the learning rate for each parameter based on past gradients. It initializes an accumulator G to store the sum of squared gradients for each parameter, which helps scale learning rates adaptively. During each update, the algorithm divides the gradient by the square root of its accumulated past values, preventing overly large updates and improving convergence in sparse data settings. The small constant ϵ ensures numerical stability, avoiding division by zero.

12.4.2   RMSprop: Root Mean Square Propagation

RMSprop improves upon AdaGrad by using an exponentially decaying average of squared gradients [6–8]:

$$E\left[g^2\right]t = \beta E\left[g^2\right]t-1 + \left(1-\beta\right)\left(g_t\right)^2,$$

$$\theta_{t+1} = \theta_t - \frac{\eta}{\sqrt{E\left[g^2\right]_t} + \epsilon} g_t,$$

where β is the decay rate, typically set to 0.9.

This modification prevents the learning rate from decreasing too quickly, allowing RMSprop to continue making progress even after many updates.

12.4.3   Adam: Adaptive Moment Estimation

Adam combines ideas from RMSprop and momentum optimization [7, 9]. It maintains both a decaying average of past gradients and past squared gradients:

First moment (momentum):

$$m_t = \beta_1 m_{t-1} + \left(1 - \beta_1\right) g_t.$$

Second moment (RMSprop):

$$v_t = \beta_2 v_{t-1} + \left(1 - \beta_2\right) g_t^2.$$

Bias correction:

$$\hat{m}_t = \frac{m_t}{1 - \beta_1^t},$$

$$\hat{v}_t = \frac{v_t}{1 - \beta_2^t};$$

Parameter update:

$$\theta_{t+1} = \theta_t - \eta \frac{\hat{m}_t}{\sqrt{\hat{v}_t} + \epsilon}.$$

Implementation example:

```python
class Adam:
def __init__(self, params, learning_rate=0.001,
            beta1=0.9, beta2=0.999, epsilon=1e-8):
    self.params = params
    self.lr = learning_rate
    self.beta1 = beta1
    self.beta2 = beta2
    self.epsilon = epsilon
    self.m = {param: np.zeros_like(param) for param in params}
    self.v = {param: np.zeros_like(param) for param in params}
    self.t = 0
def step(self, grads):
    self.t += 1
```

```python
for param, grad in zip(self.params, grads):
    # Update momentum
    self.m[param] = self.beta1 * self.m[param] + \
                    (1 - self.beta1) * grad

    # Update velocity
    self.v[param] = self.beta2 * self.v[param] + \
                    (1 - self.beta2) * (grad ** 2)

    # Bias correction
    m_hat = self.m[param] / (1 - self.beta1 ** self.t)
    v_hat = self.v[param] / (1 - self.beta2 ** self.t)

    # Update parameters
    param -= self.lr * m_hat / (np.sqrt(v_hat) + self.
epsilon)
```

This class implements Adam, that combines momentum and adaptive learning rates to improve convergence. It maintains two moving averages: momentum (m), which tracks the first moment (mean) of the gradients, and velocity (v), which tracks the second moment (variance) of the gradients. These terms are corrected for bias in early iterations to ensure stable updates. The algorithm then updates each parameter using a learning rate scaled by the corrected estimates, leading to faster convergence and better stability, especially in deep learning tasks. The small constant ϵ prevents division by zero.

12.4.4   Practical Guidelines

The choice of the optimization algorithm depends on problem nature, computational resources available, and sensitivity to the hyperparameters. Some main factors that make a difference in choosing optimizers are provided in Table 12.5.

The optimizer chosen greatly influences training efficiency and model performance:

- AdaGrad is quite suitable for those problems where data sparsity is observed; it adapts the learning rates for each of the parameters based on the historical gradient.
- Adam and RMSprop are preferable in the case of non-stationary objectives, which include deep learning tasks where gradient distributions change over time.
- Basic SGD remains a very effective choice for simple convex problems, and is computationally cheap.

Computational constraints also play an important role: while adaptive optimizers like Adam require additional memory to store the momentum and variance terms, they converge in fewer iterations; on the other hand, SGD is memory-efficient but

Table 12.5 Choosing the right optimizer: key considerations

Factor	Condition	Recommended optimizer	Remarks
Problem Characteristics	Sparse data	AdaGrad	Adjusts learning rates per parameter, making it suitable for sparse features
	Non-stationary objectives	Adam, RMSprop	Handles dynamic changes in gradient distributions effectively
	Simple convex problems	Basic SGD	Works efficiently when gradients are stable
Computational Resources	Limited memory	Basic SGD	Uses minimal memory compared to adaptive methods
	Need for fast convergence	Adam	Efficient for deep learning due to momentum and adaptive scaling
Hyperparameter Sensitivity	Default settings work well	Adam	Performs robustly with standard hyperparameters
	Requires careful tuning of decay rate	RMSprop	Sensitive to the choice of decay parameter
	Learning rate adjustment needed	AdaGrad	Accumulates past gradients, sometimes requiring a smaller initial learning rate

may only converge slower without additional techniques, such as learning rate scheduling or momentum.

Finally, there is hyperparameter sensitivity. Adam works fine with default settings in most cases, while RMSprop and AdaGrad require serious tuning of decay and learning rates. These are the trade-offs that will help you select the right optimization strategy for the problem at hand.

12.4.5   Performance Characteristics

Different optimization algorithms exhibit distinct behaviors in terms of convergence speed, generalization ability, and stability. A summary of the characteristics in Fig. 12.5 assists in choosing the appropriate optimizer.

Convergence speed varies significantly among optimizers:

- Adam is generally the fastest due to its adaptive learning rate and momentum,
- RMSprop also converges quickly but can be less stable.
- AdaGrad, on the other hand, slows down over time because its accumulated gradient term continually reduces step sizes.

Another important factor is the generalization ability:

- SGD often generalizes better because of its stochastic nature, which prevents overfitting.

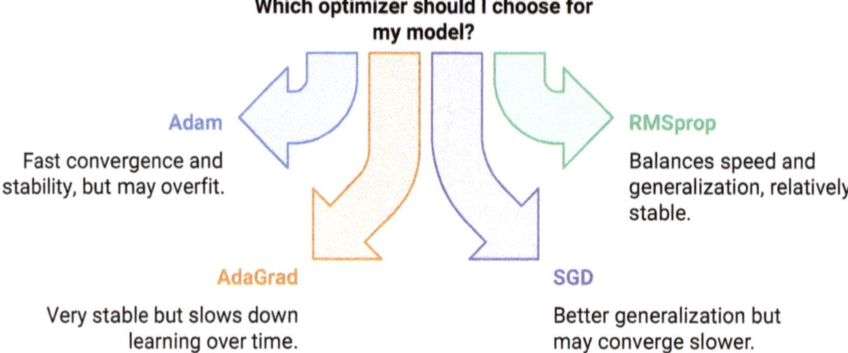

Fig. 12.5 Comparison of optimizers: convergence, generalization, and stability

- Adam, while efficient, might sometimes cause overfitting, especially if one fails to regularize it properly.
- RMSprop provides a balance between fast convergence and generalization performance.

And lastly, stability:

- Most consistently reliable, Adam dynamically adapts learning rates and implements momentum.
- RMSprop is stable if an appropriate learning rate is chosen.
- AdaGrad is inherently stable but may stop learning too early because of its aggressive learning rate reduction.

These characteristics will help one pick the best-suited optimizer for any problem, making a trade-off between speed, stability, and generalization.

12.5   Practical Applications

In this section, the real-world applications of gradient-based optimization in modern machine learning and artificial intelligence are explored. We will discuss common scenarios, practical challenges, and effective solutions.

12.5.1   Training Neural Networks

The process of training a neural network includes special optimization challenges [7, 10]: the high dimensionality of the parameter space and the non-convexity of the loss landscape.

The goal during the training in a neural network is to optimize the loss function's estimator by minimizing it:

$$L(\theta) = \frac{1}{N} \sum_{i=1}^{N} l\left(f_\theta(x_i), y_i\right),$$

where f_θ represents the neural network with parameters θ, and l is the loss function for a single example.

Common challenges in neural network optimization include the following:

1. Vanishing and Exploding Gradients. Deep structure of neural networks can result in extremely small or large gradients during backpropagation. Modern solutions include:

   - Careful initialization of weights;
   - Batch normalization between layers;
   - Residual connections in deep networks;
   - Gradient Clipping for stability.

2. Saddle Points. In high-dimensional spaces, saddle points outnumber local minima. Modern optimizers like Adam can easily escape the saddle points due to the built-in adaptive learning rate and momentum.

12.5.2   Loss Function Optimization

Different machine learning tasks require different loss functions, each presenting unique optimization challenges [11, 12]:

1. Classification Tasks The cross-entropy loss is commonly used:

$$L = -\sum_{i=1}^{N} \sum_{j=1}^{C} y_{ij} \log\left(p_{ij}\right).$$

Key considerations include:

   - Numerical stability through log-sum-exp tricks;
   - Class imbalance handling;
   - Proper normalization of logits.

2. Regression Tasks Mean squared error is standard:

$$L = \frac{1}{N} \sum_{i=1}^{N} \left(y_i - \hat{y}_i\right)^2.$$

Important aspects include:

- Feature scaling;
- Outlier handling;
- Error metric selection.

12.5.3   Optimization in Computer Vision

Computer vision tasks present specific optimization challenges [13]:

1. Convolutional Neural Networks CNNs require careful optimization strategies:

   - Proper learning rate scheduling
   - Multi-scale feature handling
   - Balanced mini-batch composition

2. Object Detection Object detection models often involve complex loss functions:

$$L = L_{loc} + \lambda L_{conf},$$

where L_{loc} represents localization loss and L_{conf} represents confidence loss.

12.5.4   Applications to Natural Language Processing

NLP models introduce some special optimization problems:

1. Transformer Architecture Training Transformer models require special optimization strategies:

   - Learning rate warmup;
   - Layer-wise learning rate decay;
   - Gradient accumulation for large models.

2. Training of Word Embedding The optimization in word embedding very often employs some specialized techniques:

   - Negative sampling;
   - Hierarchical softmax;
   - Subword tokenization.

12.5.5   Regularization in Practice

Effective regularization is critical to good optimization results [2, 3]:

1. L1 and L2 Regularization: The objective now becomes:

$$L_{reg}\left(\theta\right) = L\left(\theta\right) + \lambda_1 \left|\theta\right|_1 + \lambda_2 \left|\theta\right|_2^2.$$

2. Dropout: Applied during training with probability p:

$$y = f\left(Wx\right) \cdot \mathrm{mask}\left(p\right).$$

3. Early Stopping Monitoring validation performance to avoid overfitting while optimizing.

12.5.6 Practical Considerations for Optimization in Machine Learning

Real-world deployment of machine learning models needs to pay close attention to computational constraints, data characteristics, and deployment challenges. Table 12.6 summarizes the important factors affecting the optimization strategies in a production environment.

12.5.7 Best Practices for Effective Optimization

The following best practices should be considered in any machine learning workflow to ensure that optimization is performed effectively and efficiently in a stable manner:

Table 12.6 Practical considerations for optimization

Aspect	Challenges	Considerations
Resource Constraints	Memory limitations	Choose memory-efficient optimizers (e.g., SGD for large models)
	Computational budget	Balance optimizer complexity and training speed
	Training time constraints	Use parallelized mini-batch processing or efficient optimizers (e.g., Adam)
Data Characteristics	Handling missing values	Implement data imputation or robust preprocessing methods
	Dealing with noisy labels	Apply label smoothing or use loss functions resilient to noise
	Processing imbalanced datasets	Use weighted loss functions or data resampling techniques
Model Deployment	Quantization effects	Optimize model precision for hardware compatibility
	Inference speed requirements	Choose optimizers that enhance model efficiency (e.g., pruning, distillation)
	Model size constraints	Use compression techniques to reduce storage and memory usage

1. Model Initialization and Data Preprocessing. Using proper weight initialization methods encourages convergence. Feature normalization helps avoid the vanishing or exploding gradients problem. Data augmentation strategies will help improve model generalization.
2. Monitoring and Debugging: Learning curve analysis offers insight into the convergence behavior, while tracking the norms of gradients allows to detect vanishing or exploding gradients. Visualization of the loss landscape may give insight into optimization bottlenecks in order to provide diagnoses of issues in training.
3. Hyperparameter Selection: Learning rate scheduling, such as cosine annealing or adaptive decay, optimizes training dynamics. Batch size selection affects convergence speed and generalization, and fine-tuning optimizer parameters improves overall stability and efficiency.

By integrating these principles, machine learning practitioners can solve real-world optimization challenges and create models that are computationally efficient for production.

12.6 Advanced Topics and Future Directions

This section looks at the latest research trends and emerging developments in optimization for artificial intelligence and machine learning. We examine both theoretical advances and practical innovations that are shaping the future of the field.

12.6.1 Recent Developments in Optimization

The field of optimization continues to evolve rapidly with several promising directions:

1. Adaptive Optimization. Methods Recent research has produced several innovations in adaptive optimization. The Lion optimizer, introduced in 2023 [10], demonstrates improved convergence properties compared to Adam in many scenarios. It adapts the update rule based on both the sign and magnitude of gradients:

$$\theta_{t+1} = \theta_t - \eta \cdot sign(m_t) \cdot \text{abs}(v_t),$$

where m_t and v_t represent evolving estimates of gradient statistics.
2. Second-Order Methods. Modern implementations of second-order optimization methods are becoming more practical. The Fisher Information Matrix and Natural Gradient methods now show promise in large-scale applications [14, 15]:

$$\theta_{t+1} = \theta_t - \eta F^{-1}(\theta_t)\nabla L(\theta_t),$$

where $F(\theta_t)$ represents the Fisher Information Matrix.

12.6.2   Optimization for Large Language Models

Training large language models presents unique optimization challenges:

1. Memory-Efficient Training. Zero Redundancy Optimizer (ZeRO) and its variants enable training of larger models through sophisticated memory management and optimization strategies [16].
2. Mixed-Precision Training. Modern training combines different numerical precisions [12]: FP16 for forward/backward passes; FP32 for parameter updates.

12.6.3   Emerging Research Directions

Several promising research directions are currently being explored:

1. Quantum-Inspired Optimization. Quantum computing concepts are inspiring new optimization algorithms. These methods attempt to leverage quantum principles in classical computing environments:

   - Quantum-inspired annealing [13, 17];
   - Quantum-inspired genetic algorithms [18–20];
   - Hybrid quantum-classical optimization [21].

2. Neural Architecture Optimization. The automated discovery of optimal neural network architectures represents a meta-optimization problem:

$$\min_{\alpha \in A} \min_{\theta \in \Theta} L(\alpha,\theta),$$

where α represents architecture parameters and θ represents model weights.

12.6.4   Outlook

Optimization research continues to evolve rapidly. Future developments will likely focus on both theoretical and practical advancements. Deeper mathematical insights into non-convex optimization, stochastic processes, and information geometry will strengthen foundational understanding.

From a practical perspective, improvements in large-scale training, resource efficiency, and robustness will drive more effective optimization techniques.

Additionally, new paradigms such as neuromorphic optimization, biologically inspired methods, and quantum-based approaches could redefine optimization strategies.

As AI models grow in complexity, optimization will play an increasingly crucial role in ensuring efficiency, scalability, and sustainability. Ongoing research in this field promises to unlock new capabilities, enabling more powerful and adaptable machine learning systems.

12.7 Practice and Exercises

12.7.1 Practice and Programming Exercises

All programming exercises for this chapter are available in the Google Colab notebook. You can access the complete notebook by scanning the QR code below or using this link: https://colab.research.google.com/drive/1GHT6ksheMFJj-Svo0Bk1RxWUwX-0esYO

The notebook contains interactive Python code that you can run and modify to better understand gradient-based optimization concepts. Each exercise builds upon the previous ones, gradually introducing more complex aspects of optimization algorithms. Solutions to the exercises are available to instructors through the publisher's resources.

Exercise 1: Basic Gradient Descent Implementation (Notebook Cells 1–3)

Implement gradient descent for linear regression with two parameters. This exercise demonstrates the fundamental concepts of gradient-based optimization.

Tasks:
- Generate synthetic data with known parameters;
- Implement basic gradient descent algorithm;

...ze the results;
Analyze the optimized parameters.

Learning Objectives:
- Understanding the basic gradient descent algorithm;
- Implementing parameter updates;
- Visualizing optimization results.

Exercise 2: Cost Function Analysis (Notebook Cells 4–6)

Explore the behavior of the cost function during optimization.

Tasks:
- Implement the MSE cost function;
- Track cost during optimization;
- Visualize cost evolution on a logarithmic scale;
- Analyze convergence behavior.

Learning Objectives:
- Understanding cost function behavior;
- Monitoring optimization progress;
- Interpreting convergence patterns.

Exercise 3: Impact of Initialization (Notebook Cells 7–9)

Investigate how different parameter initializations affect optimization.

Tasks:
- Implement gradient descent with multiple initializations;
- Compare convergence paths;
- Analyze the effect on final solutions;
- Visualize optimization trajectories.

Learning Objectives:
- Understanding initialization importance;
- Analyzing convergence stability;
- Comparing optimization paths.

Exercise 4: Comparing Optimization Algorithms (Notebook Cells 10–13)

Compare Batch Gradient Descent (BGD) and Stochastic Gradient Descent (SGD).

Tasks:
- Implement both BGD and SGD;
- Compare convergence behavior;

- Analyze computational efficiency;
- Visualize learning dynamics.

Learning Objectives:
- Understanding different optimization variants;
- Analyzing trade-offs between methods;
- Implementing stochastic updates.

Exercise 5: Neural Network Optimization (Notebook Cells 14–17)

Apply optimization algorithms to neural network training.

Tasks:
- Implement a simple neural network;
- Compare SGD and Adam optimizers;
- Analyze training dynamics;
- Visualize learning curves.

Learning Objectives:
- Understanding neural network optimization;
- Implementing advanced optimizers;
- Analyzing training behavior.

Note: Students should refer to the lecture materials and textbook sections covered in this chapter while working through these exercises. The accompanying Python notebook provides additional guidance and helper functions.

Programming Guidelines

When working with the exercises:

- Run the cells in order;
- Experiment with different hyperparameters;
- Analyze the impact of your changes;
- Document your observations.

Additional Challenges

For each exercise, consider these extensions:

- Modify the learning rate and analyze its impact;
- Add momentum to the optimization;
- Implement early stopping;
- Experiment with different data distributions.

should:

- … on plots;
- … results;
- Analyze convergence behavior;
- Document your findings.

Discussion Questions

After completing the exercises, consider:

1. How do different optimizers compare in terms of:

   - Convergence speed;
   - Stability;
   - Final solution quality.

2. What role does initialization play?
3. How do different hyperparameters affect optimization?
4. What are the practical trade-offs between methods?

12.7.2   Self-Assessment Questions

Instructions for Self-Assessment

Before attempting the questions:

- Review the chapter material thoroughly;
- Complete all practical exercises;
- Understand both theoretical concepts and their practical applications.

For each question:

- Read carefully and consider all options;
- Try to solve without referring to materials;
- Check your answers with the provided key;
- Review explanations for any incorrect answers.

Multiple Choice Questions

1. What best describes the role of the learning rate in gradient descent?

   (a) Determines the final accuracy of the model
   (b) Controls the size of steps taken during optimization

   (c) Defines the number of iterations required
   (d) Specifies the batch size for training

2. In the context of gradient-based optimization, what is a saddle point?

   (a) A point of minimum local gradient
   (b) A point where the gradient is zero but not an extremum
   (c) The global minimum of the loss function
   (d) A point where the function is undefined

3. Why is stochastic gradient descent often preferred over batch gradient descent?

   (a) It always finds the global minimum
   (b) It requires less memory and updates more frequently
   (c) It converges in fewer total iterations
   (d) It provides more stable gradients

4. What is the main advantage of the Adam optimizer?

   (a) It always finds the global minimum
   (b) It combines momentum and adaptive learning rates
   (c) It requires no hyperparameter tuning
   (d) It guarantees convergence in all cases

5. Which statement about mini-batch gradient descent is correct?

   (a) It's always slower than stochastic gradient descent
   (b) It provides a compromise between batch and stochastic methods
   (c) It requires more memory than batch gradient descent
   (d) It always converges to the global minimum

6. What is the primary purpose of momentum in gradient descent?

   (a) To increase the learning rate over time
   (b) To accumulate past gradients and accelerate convergence
   (c) To prevent overfitting during training
   (d) To guarantee finding the global minimum

7. How does AdaGrad differ from basic gradient descent?

   (a) It uses a fixed learning rate
   (b) It adapts the learning rate for each parameter
   (c) It eliminates the need for gradients
   (d) It only works with neural networks

8. What challenge does the vanishing gradient problem present?

   (a) Learning becomes too fast in deep layers
   (b) Gradients become too small to enable effective learning
   (c) Memory requirements increase exponentially
   (d) The model always overfits

9. When implementing gradient descent, what is the most appropriate initial learning rate?

    (a) Always use 0.01
    (b) Use the largest value that maintains stable convergence
    (c) The same as the batch size
    (d) Any random value between 0 and 1

10. Which technique helps prevent overshooting in gradient descent?

    (a) Increasing the learning rate
    (b) Using adaptive learning rates
    (c) Removing momentum terms
    (d) Maximizing batch size

Answer Key and Explanations

1. (b) Controls the size of steps taken during optimization

   - The learning rate determines how large each update step is during optimization
   - Too large: may overshoot; too small: slow convergence

2. (b) A point where the gradient is zero but not an extremum

   - Saddle points have zero gradient but are neither minima nor maxima
   - Common in high-dimensional optimization problems

3. (b) It requires less memory and updates more frequently

   - Processes one example at a time, enabling frequent updates
   - Particularly useful for large datasets
   - Can escape local minima more easily due to noise in gradients

4. (b) It combines momentum and adaptive learning rates

   - Incorporates both first and second moments of gradients
   - Adapts learning rates for each parameter
   - Often works well with default hyperparameters

5. (b) It provides a compromise between batch and stochastic methods

   - Balances computational efficiency with stability
   - Enables efficient use of modern hardware
   - Provides more stable convergence than SGD

6. (b) To accumulate past gradients and accelerate convergence

   - Helps overcome local minima and saddle points
   - Reduces oscillations in high-curvature regions
   - Speeds up convergence in consistent directions

7. (b) It adapts the learning rate for each parameter

   - Maintains separate learning rates for each parameter
   - Reduces learning rates for frequently updated parameters
   - Helps handle different scales in parameters

8. (b) Gradients become too small to enable effective learning

   - Common problem in deep networks
   - Makes training deep layers difficult
   - Can be addressed with proper initialization and normalization

9. (b) Use the largest value that maintains stable convergence

   - Requires experimentation for each problem
   - Should be adjusted based on observed convergence
   - Can be reduced during training if needed

10. (b) Using adaptive learning rates

   - Prevents overshooting by adjusting step sizes
   - Accounts for gradient magnitude
   - Provides more stable convergence

12.8   Conclusion

Gradient-based optimization forms the foundation of modern machine learning and artificial intelligence. This chapter has explored the theoretical principles, practical implementations, and application considerations for these powerful methods.

The fundamental gradient descent algorithm provides a simple yet effective approach for minimizing differentiable functions. Through iterative steps in the direction opposite to the gradient, these methods navigate complex optimization landscapes. The introduction of stochastic variations and mini-batch processing addresses computational efficiency challenges for large datasets.

Advanced methods like momentum, AdaGrad, RMSprop, and Adam enhance performance through adaptive learning rates and historical gradient information. These techniques significantly improve convergence speed and stability, particularly in non-convex problems common in deep learning.

The choice of optimization algorithm substantially impacts model performance. Selecting appropriate methods, tuning hyperparameters, and monitoring convergence represent essential skills for building efficient AI systems. The future of optimization includes advanced techniques like second-order methods, quantum-inspired approaches, and specialized algorithms for large-scale models.

References

1. Tran-Dinh, Q., van Dijk, M.: Chapter 1—Gradient descent-type methods: Background and simple unified convergence analysis. In: Nguyen, L.M., Hoang, T.N., and Chen, P.-Y. (eds.) Federated Learning. pp. 3–28. Academic Press (2024). https://doi.org/10.1016/B978-0-44-319037-7.00008-9.
2. Zakwan, M.: Chapter 14—Gradient-based optimization. In: Eslamian, S. and Eslamian, F. (eds.) Handbook of Hydroinformatics. pp. 243–251. Elsevier (2023). https://doi.org/10.1016/B978-0-12-821285-1.00013-0.
3. Brereton, R.G.: Optimisation: Steepest Ascent, Steepest Descent and Gradient Methods☆. In: Brown, S., Tauler, R., and Walczak, B. (eds.) Comprehensive Chemometrics (Second Edition). pp. 573–583. Elsevier, Oxford (2020). https://doi.org/10.1016/B978-0-12-409547-2.14835-8.
4. Hong, Y., Lin, J.: High probability bounds on AdaGrad for constrained weakly convex optimization. Journal of Complexity. 86, 101889 (2025). https://doi.org/10.1016/j.jco.2024.101889.
5. Traoré, C., Pauwels, E.: Sequential convergence of AdaGrad algorithm for smooth convex optimization. Operations Research Letters. 49, 452–458 (2021). https://doi.org/10.1016/j.orl.2021.04.011.
6. Shen, L., Chen, C., Zou, F., Jie, Z., Sun, J., Liu, W.: A Unified Analysis of AdaGrad With Weighted Aggregation and Momentum Acceleration. IEEE Transactions on Neural Networks and Learning Systems. 35, 14482–14490 (2024). https://doi.org/10.1109/TNNLS.2023.3279381.
7. Xue, M., Li, J., Luo, Q.: Toward Optimal Learning Rate Schedule in Scene Classification Network. IEEE Geoscience and Remote Sensing Letters. 19, 1–5 (2022). https://doi.org/10.1109/LGRS.2020.3040359.
8. Xu, D., Zhang, S., Zhang, H., Mandic, D.P.: Convergence of the RMSProp deep learning method with penalty for nonconvex optimization. Neural Networks. 139, 17–23 (2021). https://doi.org/10.1016/j.neunet.2021.02.011.
9. Chen, C., Shen, L., Liu, W., Luo, Z.-Q.: Efficient-Adam: Communication-Efficient Distributed Adam. IEEE Transactions on Signal Processing. 71, 3257–3266 (2023). https://doi.org/10.1109/TSP.2023.3309461.
10. Chen, X., Liang, C., Huang, D., Real, E., Wang, K., Pham, H., Dong, X., Luong, T., Hsieh, C.-J., Lu, Y., Le, Q.V.: Symbolic discovery of optimization algorithms. In: Proceedings of the 37th International Conference on Neural Information Processing Systems. pp. 49205–49233. Curran Associates Inc., Red Hook, NY, USA (2024).
11. Lei, Y.-J., Wang, B.-Y., Yang, Y.-T.: Optimizing the loss function for bounding box regression through scale smoothing. Ain Shams Engineering Journal. 15, 103046 (2024). https://doi.org/10.1016/j.asej.2024.103046.
12. Micikevicius, P., Narang, S., Alben, J., Diamos, G., Elsen, E., Garcia, D., Ginsburg, B., Houston, M., Kuchaiev, O., Venkatesh, G., Wu, H.: Mixed Precision Training, http://arxiv.org/abs/1710.03740, (2018). https://doi.org/10.48550/arXiv.1710.03740.
13. Heidari, S., Dinneen, M.J., Delmas, P.: Quantum Annealing for Computer Vision minimization problems. Future Generation Computer Systems. 160, 54–64 (2024). https://doi.org/10.1016/j.future.2024.05.037.
14. Jaćimović, V.: Natural gradient ascent in evolutionary games. Biosystems. 236, 105127 (2024). https://doi.org/10.1016/j.biosystems.2024.105127.
15. Zhu, J.-X., Zhu, Z., Au, S.-K.: Accelerating computations in two-stage Bayesian system identification with Fisher information matrix and eigenvalue sensitivity. Mechanical Systems and Signal Processing. 186, 109843 (2023). https://doi.org/10.1016/j.ymssp.2022.109843.
16. Rajbhandari, S., Rasley, J., Ruwase, O., He, Y.: ZeRO: memory optimizations toward training trillion parameter models. In: Proceedings of the International Conference for High Performance Computing, Networking, Storage and Analysis. pp. 1–16. IEEE Press, Atlanta, Georgia (2020).

17. Salloum, H., Salloum, A., Mazzara, M., Zykov, S.: Quantum Annealing in Machine Learning: QBoost on D-Wave Quantum Annealer. Procedia Computer Science. 246, 3285–3293 (2024). https://doi.org/10.1016/j.procs.2024.09.311.
18. Guo, J., Sun, L., Wang, R., Yu, Z.: An Improved Quantum Genetic Algorithm. In: 2009 Third International Conference on Genetic and Evolutionary Computing. pp. 14–18 (2009). https://doi.org/10.1109/WGEC.2009.41.
19. Peng, W.: Quantum Model of Genetic Algorithm. In: 2008 International Symposium on Knowledge Acquisition and Modeling. pp. 576–580 (2008). https://doi.org/10.1109/KAM.2008.11.
20. Hussain, M., Wei, L.-F., Rehman, A., Ali, M., Waqas, S.M., Abbas, F.: Cost-aware quantum-inspired genetic algorithm for workflow scheduling in hybrid clouds. Journal of Parallel and Distributed Computing. 191, 104920 (2024). https://doi.org/10.1016/j.jpdc.2024.104920.
21. Ellinas, P., Chevalier, S., Chatzivasileiadis, S.: A hybrid quantum–classical algorithm for mixed-integer optimization in power systems. Electric Power Systems Research. 235, 110835 (2024). https://doi.org/10.1016/j.epsr.2024.110835.

Chapter 13
Tabu Search

Abstract This chapter examines Tabu Search (TS), a powerful metaheuristic that enhances local search with adaptive memory structures. TS overcomes limitations of simpler optimization methods by using memory to avoid cycling and escape local optima. The chapter begins with fundamental concepts of short-term memory and tabu lists. It then explores advanced features including aspiration criteria, intensification strategies, and diversification mechanisms. Implementation details focus on efficient data structures and practical applications in vehicle routing, job scheduling, and network design. Python code examples demonstrate both basic implementations.

13.1 Foundations of Tabu Search

Tabu Search (TS) represents a significant advancement in metaheuristic optimization methods [1, 4]. Unlike simpler approaches like Hill Climbing or basic local search, TS introduces sophisticated memory structures to guide the search process through complex solution spaces [5, 6]. This section establishes the fundamental concepts and principles that make TS a powerful tool for solving combinatorial optimization problems.

13.1.1 Basic Concepts and Motivation

At its core, Tabu Search addresses a fundamental limitation of local search methods: their tendency to become trapped in local optima [1, 4]. Consider a minimization problem formally defined as:

$$\min_{x \in X} f(x),$$

O. Kuznetsov, *Intelligent Systems: From Theory to Applications*, Cognitive Technologies, https://doi.org/10.1007/978-3-032-00044-6_13

where:

- X represents the feasible solution space;
- $f : X \to R$ is the objective function to minimize;

The key innovation of Tabu Search lies in its systematic use of memory to [2, 3]:

- Prevent cycling back to recently visited solutions;
- Guide the search toward promising regions of the solution space;
- Allow escape from local optima through controlled acceptance of non-improving moves.

The basic TS mechanism operates by maintaining a tabu list T that prohibits certain moves:

$$x_{t+1} = \arg\min_{x \in N(x_t) \setminus T(t)} f(x),$$

where:

- $N(x_t)$ is the neighborhood of the current solution;
- $T(t)$ represents the set of tabu moves at iteration t.

13.1.2 Historical Development

Tabu Search emerged from the work of Fred Glover in 1986 as a response to limitations of traditional optimization methods [4]. The development of TS can be traced through several key stages (Fig. 13.1).

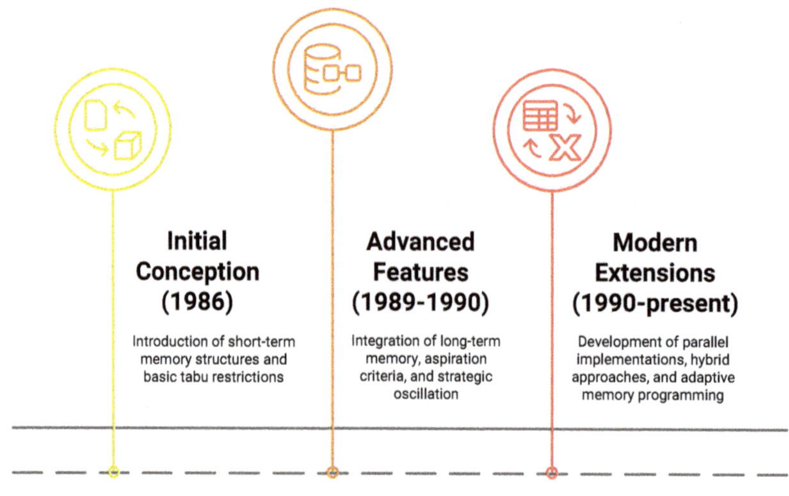

Fig. 13.1 Development stages of Tabu Search

The embryonic version of Tabu Search was born in 1986 and introduced basic mechanisms such as short-term memory structures and elementary tabu restrictions for conducting the search [4]. Throughout the late 1980s and early 1990s, the approach further developed with the incorporation of long-term memory, allowing a more advanced exploration of the solution space [2, 3]. Other key developments during this period included aspiration criteria and strategic oscillation, which enhanced the capability of the algorithm to escape from local optima. Other developments since 1990 have included modern extensions such as parallel implementations to enhance computational efficiency, hybrid approaches that combine Tabu Search with other optimization techniques, and adaptive memory programming, which dynamically adjusts memory structures to enhance performance [7, 8].

13.1.3 Comparison with Other Search Methods

Tabu Search differs from other metaheuristics in several important ways:

1. Versus Hill Climbing:

   - Hill Climbing:

   $$x_{t+1} = \arg\min_{x \in N(x_t)} f(x);$$

   - Tabu Search:

   $$x_{t+1} = \arg\min_{x \in N(x_t) \setminus T(t)} f(x);$$

2. Versus Simulated Annealing:

   - SA uses probabilistic acceptance:

   $$P(\text{accept}) = e^{-\Delta E / T};$$

   - TS uses deterministic memory-based decisions;

3. Versus Genetic Algorithms:

   - GAs maintain a population of solutions;
   - TS works with a single solution trajectory but maintains memory of search history.

Tabu search has a variety of important advantageous properties, as outlined in Fig. 13.2. The method explicitly utilizes search history to guide decision-making, allowing the algorithm to systematically explore the solution space. A crucial feature is its ability to escape local optima by permitting controlled non-improving moves, preventing premature convergence. Besides, Tabu Search keeps a good balance between intensification and diversification, ensuring both focused exploration

Fig. 13.2 Factors contributing to Tabu Search effectiveness

of the most promising regions and broad coverage of the solution space. Its deterministic nature enhances reproducibility further, hence being one of the most reliable optimization techniques.

The core strength of Tabu Search comes through its adaptive memory principles that structure the search process across different time scales. Short-term memory avoids cycling by preventing recently visited solutions from being revisited-thus assuring some form of making progress. The intermediate-term memory supports intensification in a manner reinforced by making those moves leading to high-quality solutions, while long-term memory promotes diversification in order to explore previously underexplored regions.

This memory-based approach makes Tabu Search particularly effective when dealing with complex combinatorial problems on which the traditional heuristics obtain poor results since they are computationally complex. It is perfectly suited for problems marked by the existence of several local optima, as it systematically tours the space in search of answers. Besides, capability for balancing exploitation and exploration of knowledge enhances optimization performance in tasks strategically making decisions. Finally, its tractability for practical applications with complex constraints makes it very powerful in logistics, scheduling, network design, and even more generally in artificial intelligence.

Understanding these foundational concepts is crucial for effectively implementing and applying Tabu Search to practical optimization problems, which we will explore in detail in subsequent sections.

13.2   Core Components and Mechanisms

The effectiveness of Tabu Search stems from its carefully designed core components. Each component plays a specific role in guiding the search process through complex solution spaces. This section explores these essential mechanisms and their interactions.

13.2.1   Short-Term Memory (Tabu List)

The tabu list is the primary mechanism for avoiding cycles in the search process. At time t, we can represent the tabu list as [2, 3]:

$$T(t) = m_{t-k+1}, m_{t-k+2}, \ldots, m_t,$$

where:

- m_i represents a move or solution attribute made tabu;
- k is the tabu tenure (length of the list).

The tabu list affects the search process by modifying the available moves:

- Available moves = Neighborhood moves − Tabu moves;
- Formally:

$$N'(x_t) = N(x_t) \setminus T(t).$$

Key design considerations for the tabu list include:

1. What to make tabu (moves, attributes, or solutions);
2. How long to maintain tabu status (fixed or dynamic tenure);
3. How to efficiently check tabu status.

13.2.2   Aspiration Criteria

Aspiration criteria allow overriding tabu status when certain conditions are met. The most common criterion is improvement over the best known solution:

$$\text{Accept move } m \text{ if } f(x \oplus m) < f(x^*),$$

where:

- x^* is the best solution found so far;
- $\oplus$ denotes the application of move m to solution x.

Common aspiration criteria types:

1. Global aspiration: Better than best known solution;
2. Recent aspiration: Better than recent solutions;
3. Move-specific aspiration: Based on move characteristics.

13.2.3   Move Selection and Neighborhood Structure

The neighborhood structure defines possible transitions between solutions. For a current solution x, its neighborhood $N(x)$ consists of solutions reachable by a single move:

$$N(x) = x' \in X : x' = x \oplus m, m \in M(x),$$

where $M(x)$ is the set of possible moves from x.

Move selection follows these steps:

1. Generate candidate moves: $M(x_t)$;
2. Filter tabu moves:

$$M'(x_t) = M(x_t) \setminus T(t);$$

3. Select best non-tabu move:

$$m^* = \arg\min_{m \in M'(x_t)} f(x_t \oplus m).$$

13.2.4   Long-Term Memory Structures

Long-term memory enhances the search process through:

1. Frequency-based memory
2. Quality-based memory
3. Recency-based memory

For frequency-based memory, we maintain:

$$\text{freq}(i) = \frac{\text{occurrences of feature } i}{\text{total iterations}}.$$

This information modifies the objective function:

$$f'(x) = f(x) + \lambda \sum_i \frac{\text{freq}(i)}{\text{total}_i \text{ter}},$$

where:

- λ is a weighting parameter;
- The sum is over solution features.

Long-term memory in Tabu Search relies on some crucial mechanisms to enhance the exploration of the solution space and orient the search process, as described in Fig. 13.3. This includes recording solution attributes: residence frequency of solutions to know which areas have been visited the most, transition frequency to identify characteristics of movement, and solution quality related to specific properties in order to recognize the most promising regions in which the search should focus.

This information is then actively used to refine the search strategy once recorded. The recorded data can change the objective function by embedding historical knowledge into the decision-making process. It also plays a vital role in guiding diversification to ensure that the search will explore new regions instead of repeatedly visiting known solutions. Besides, the long-term memory is used to control the intensity of the search by updating the intensity of exploration based on past performance, which may allow a balance between intensification and diversification to be modified dynamically. All these mechanisms make use of the gathered knowledge in improving efficiency regarding Tabu Search in complicated optimization problems.

These components work together through:

1. Move Evaluation:

$$\text{score}(m) = f(x \oplus m) + P(m) - A(m),$$

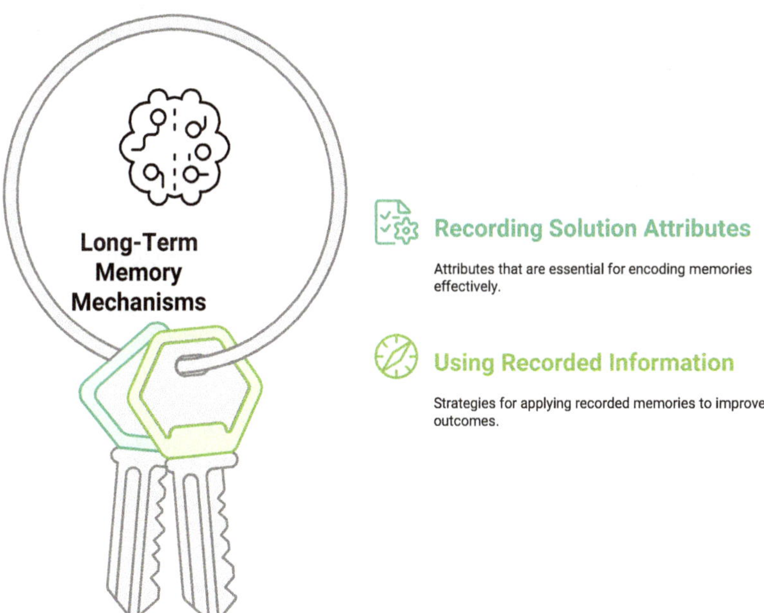

Fig. 13.3 Enhancing memory through strategic recording and application

where:

- $P(m)$ is penalty for tabu status;
- $A(m)$ is aspiration bonus.

2. Solution Selection:

$$x_{t+1} = x_t \oplus \arg\min_{m \in M'(x_t)} \text{score}(m).$$

3. Memory Update:

- Update tabu list:

$$T(t+1) = m_{t-k+2}, \ldots, m_t, m^*.$$

- Update frequency counts;
- Update best solution if improved.

The effectiveness of Tabu Search largely depends on the proper balance and interaction of these components. Understanding their roles and relationships is crucial for successful implementation and application to real-world problems.

13.3   Advanced Features

The basic Tabu Search framework can be enhanced with several advanced features that significantly improve its performance in complex optimization scenarios. This section explores these sophisticated mechanisms and their implementation.

13.3.1   Intensification Strategies

Intensification focuses the search in promising regions of the solution space. The process involves:

1. Memory-based Intensification

- Record elite solutions in a set E;
- Return to promising regions using:

$$x_{return} = \arg\min_{x \in E} f(x).$$

2. Attribute-based Intensification

- Track good solution attributes:

$$\text{score}(a) = \frac{\sum_{x \in E} \text{value}(x) \cdot \text{has}(x,a)}{\sum_{x \in E} \text{has}(x,a)},$$

- where:
  - value(x) is solution quality
  - has(x, a) is 1 if solution x has attribute a

3. Modified Objective Function

$$f'(x) = f(x) - \alpha \sum_{a \in A(x)} \text{score}(a),$$

where:

- $A(x)$ is the set of attributes in solution x
- α is an intensification parameter

13.3.2 Diversification Mechanisms

Diversification ensures broad exploration of the solution space through:

1. Frequency-based Penalties

$$f'(x) = f(x) + \beta \sum_{i \in F(x)} \frac{\text{freq}(i)}{\text{total}_i \text{ter}},$$

where:

- $F(x)$ represents features of solution x;
- β is a diversification parameter.

2. Restart Mechanisms

- After k iterations without improvement;
- Using modified initial solution:

$$x_{new} = \arg\max_{x \in X} \text{distance}(x, x_{history}).$$

3. Strategic Oscillation

- Alternate between feasible and infeasible regions;
- Adjust penalty parameters dynamically:

$$\omega_{t+1} = \begin{cases} \omega_t \cdot \gamma & \text{if feasible} \\ \omega_t / \gamma & \text{if infeasible} \end{cases}.$$

13.3.3   Strategic Oscillation

Strategic oscillation involves systematic moves across boundaries in solution space:

1. Boundary Functions

   - Define boundary measure: $b(x)$;
   - Set target levels: $[b_{min}, b_{max}]$;
   - Guide oscillation:

$$\text{target}t = bmin + \left(\frac{t \bmod T}{T}\right)(b_{\max} - b_{\min}).$$

2. Controlled Violation

$$f'(x) = f(x) + \lambda \left| b(x) - \text{target}_t \right|,$$

where λ controls oscillation strength.
3. Depth of Oscillation

   - Interior phase: Stay within boundaries;
   - Exterior phase: Allow controlled boundary violations;
   - Transition phase: Move between phases systematically.

13.3.4   Path Relinking

Path relinking connects promising solutions through structured trajectories:

1. Basic Algorithm For initiating solution x_i and guiding solution x_g:

   - Generate path: $P(x_i, x_g)$;
   - Evaluate intermediate solutions:

$$x^* = \arg\min_{x \in P(x_i, x_g)} f(x);$$

2. Path Generation

$$x_{t+1} = x_t \oplus \arg\min_{m \in M(x_t)} \text{distance}\left(x_t \oplus m, x_g\right),$$

where distance measures solution similarity;
3. Implementation Strategies

   - Forward relinking: Start from x_i;
   - Backward relinking: Start from x_g;
   - Bidirectional relinking: Meet in middle

Key Implementation Considerations:

1. Balance Control

    - Monitor intensification/diversification ratio:

    $$r = \frac{time\_in\_intensification}{total\_time};$$

    - Adjust parameters accordingly;

2. Adaptive Mechanisms

    - Update memory structures dynamically;
    - Modify search parameters based on progress:

    $$parameter_t = parameter_0 \cdot (1 + \gamma \cdot progress);$$

3. Integration of Features

    - Combine multiple mechanisms strategically;
    - Avoid conflicting patterns;
    - Maintain search coherence.

These advanced features significantly enhance the basic Tabu Search algorithm's ability to handle complex optimization problems. The key to successful implementation lies in carefully balancing these mechanisms and adapting them to specific problem characteristics.

13.4   Implementation Aspects

This section provides practical guidance for implementing Tabu Search effectively. We focus on essential implementation details and provide concrete examples using Python.

13.4.1   Basic Algorithm Structure

The fundamental Tabu Search algorithm can be implemented using this basic structure:

```python
class TabuSearch:

def
__init__(self, initial_solution, tabu_size=10, max_
iterations=100):
```

```
        self.current = initial_solution
        self.best = initial_solution
        self.tabu_list = []
        self.tabu_size = tabu_size
        self.max_iterations = max_iterations
    def search(self):
        iteration = 0
        while iteration < self.max_iterations:
            neighbors = self.get_neighbors(self.current)
            best_move = self.select_best_move(neighbors)
            self.update_solution(best_move)
            self.update_tabu_list(best_move)
            iteration += 1
        return self.best
```

Key components include:

1. Solution representation;
2. Neighborhood generation;
3. Move selection;
4. Tabu list management.

13.4.2   Data Structures and Efficiency

The choice of data structures significantly impacts performance:

1. Tabu List Implementation:

```
class TabuList:
   def __init__(self, max_size):
       self.list = collections.deque(maxlen=max_size)
   def is_tabu(self, move):
       return move in self.list
   def add(self, move):
       self.list.append(move)
```

2. Solution Representation. Consider the example of a traveling salesman problem:

```
class TSPSolution:
   def __init__(self, route):
       self.route = route
       self._cost = None
   @property
   def cost(self):
```

```
    if self._cost is None:
        self._cost = self.calculate_cost()
    return self._cost
```

3. Neighborhood Generation:

```
def get_neighbors(self, solution):
  neighbors = []
  n = len(solution.route)
  for i in range(n):
      for j in range(i+1, n):
          new_route = solution.route.copy()

new_route[i], new_route[j] = new_route[j], new_route[i]
          neighbors.append(TSPSolution(new_route))
    return neighbors
```

13.4.3 Parameter Settings

Key parameters that affect performance include:

1. Tabu Tenure. The tabu tenure τ can be:

   - Fixed:

$$\tau = k\left(\text{constant}\right);$$

   - Dynamic:

$$\tau_t = \max\left(k_{\min}, \min\left(k_{\max}, f(t)\right)\right);$$

   - Reactive: Based on solution quality;

2. Aspiration Threshold

```
def aspiration_criteria(self, move, cost):
  return cost < self.best_cost
```

3. Search Duration

   - Maximum iterations;
   - Time limit;
   - Improvement threshold.

13.4.4   *Python Implementation Examples*

1. Core Search Loop with Aspiration:

```python
def tabu_search(self):
  while not self.stopping_criterion():
      current_neighbors = self.get_neighbors()
      best_move = None
      best_cost = float('inf')

      for move in current_neighbors:
          if (not self.is_tabu(move) or
              self.aspiration_criteria(move)):
              cost = self.evaluate_move(move)
              if cost < best_cost:
                  best_move = move
                  best_cost = cost

      self.apply_move(best_move)
      self.update_tabu_list(best_move)
```

2. Memory Structures:

```python
class AdaptiveMemory:
  def __init__(self):
      self.short_term = TabuList()
      self.frequency = defaultdict(int)
      self.elite_solutions = []
  def update(self, solution, move):
      self.short_term.add(move)
      self.frequency[move] += 1
      if self.is_elite(solution):
          self.elite_solutions.append(solution)
```

3. Intensification and Diversification:

```python
def intensify(self):
  return to_best_known_region()
def diversify(self):
  return restart_from_new_region()
def adaptive_search(self):
  if self.stagnation_detected():
      if self.should_intensify():
          self.intensify()
      else:
          self.diversify()
```

Implementation Best Practices (Fig. 13.4):

1. Efficiency Considerations

    • Cache solution evaluations;
    • Use appropriate data structures;
    • Implement efficient neighborhood exploration;

2. Robustness

    • Handle edge cases;
    • Validate input parameters;
    • Implement proper error handling;

3. Modularity

    • Separate core algorithm from problem-specific code;
    • Allow easy parameter tuning;
    • Enable component reuse.

This implementation framework provides a foundation for developing efficient and effective Tabu Search algorithms. The key is to maintain a balance between code efficiency and readability while ensuring the implementation remains flexible enough to accommodate problem-specific requirements.

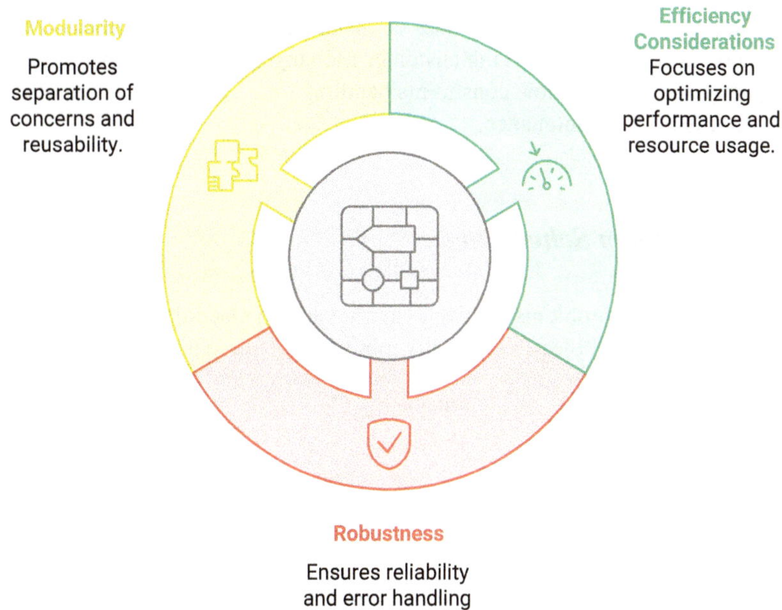

Fig. 13.4 Tabu Search implementation strategies

13.5 Applications and Case Studies

Tabu Search has proven highly effective across diverse optimization problems. This section explores practical applications, highlighting how the algorithm's features address specific challenges in different domains.

13.5.1 *Vehicle Routing Problems*

Vehicle Routing Problems (VRP) represent a classic application of Tabu Search. The basic problem involves optimizing delivery routes for a fleet of vehicles serving multiple customers. For a VRP, we aim to minimize:

$$f(x) = \sum_{k=1}^{K}\sum_{i=1}^{n_k} d\left(v_{k,i}, v_{k,i+1}\right),$$

where:

- K represents the number of vehicles;
- n_k is the number of customers served by vehicle k;
- $d(v_i, v_j)$ denotes the distance between locations.

Key implementation considerations for VRP include:

- Solution representation as sequences of customer visits;
- Neighborhood moves involving customer exchange between routes;
- Capacity and time window constraints handling;
- Route feasibility maintenance.

13.5.2 *Job Shop Scheduling*

Job shop scheduling problems involve assigning jobs to machines while minimizing completion time. The objective typically minimizes makespan:

$$\min \max_{j \in J} C_j,$$

where:

- C_j represents the completion time of job j;
- J is the set of all jobs.

Tabu Search features particularly useful for scheduling include:

- Memory structures tracking successful operation sequences;
- Neighborhood moves based on operation swaps and shifts;

- Constraint handling for operation precedence;
- Critical path analysis for move evaluation.

13.5.3   *Network Design Problems*

Network design optimization encompasses problems from telecommunications to transportation infrastructure. A typical objective function combines cost and performance:

$$f(x) = \alpha \sum_{e \in E} c_e x_e + \beta \sum_{d \in D} p_d(x),$$

where:

- c_e represents edge installation costs;
- $p_d(x)$ measures network performance;
- α, β are weighting factors.

The Tabu Search approach for network design emphasizes:

- Long-term memory for network topology patterns;
- Strategic oscillation between feasible and infeasible solutions;
- Multi-objective consideration of cost and performance;
- Incremental evaluation of network changes.

13.5.4   *Facility Location Problems*

Facility location problems involve determining optimal placement of facilities to serve customer demand. The objective typically minimizes total cost:

$$\min \sum_{i \in I} f_i y_i + \sum_{i \in I} \sum_{j \in J} c_{ij} x_{ij},$$

where:

- f_i represents facility opening costs;
- c_{ij} denotes transportation costs;
- y_i indicates if facility i is open;
- x_{ij} represents customer assignments.

Tabu Search adaptations for facility location include:

- Memory structures recording successful facility configurations;
- Compound moves involving facility opening/closing and customer reassignment;

- Strategic oscillation between different coverage levels;
- Intensification around promising facility combinations.

In real-world applications, several common themes emerge:

1. Solution Quality Assessment. The evaluation often involves multiple criteria:

$$f(x) = w_1 f_1(x) + w_2 f_2(x) + \ldots + w_k f_k(x),$$

2. Constraint Handling. Most practical problems include various constraints:

   - Hard constraints that must be satisfied;
   - Soft constraints that can be violated with penalties;
   - Dynamic constraints that change over time.

3. Computational Efficiency. Real applications require efficient implementation:

   - Incremental solution evaluation;
   - Selective neighborhood exploration;
   - Parallel processing where applicable;
   - Memory management for large-scale problems.

Each application domain presents unique challenges that Tabu Search addresses through its adaptive memory structures and flexible search strategies. Success in practical applications often depends on carefully balancing these elements while maintaining computational efficiency.

The effectiveness of Tabu Search in these diverse applications demonstrates its versatility as an optimization tool. Understanding these applications provides valuable insights for adapting the algorithm to new problem domains.

13.6   Practical Considerations and Future Directions

This section explores practical aspects of implementing Tabu Search in real-world scenarios and examines emerging trends in its development. We focus on implementation challenges, performance optimization, and future research directions.

13.6.1   Implementation Challenges

Real-world implementations of Tabu Search face several key challenges that require careful consideration.

1. Memory Management. For large-scale problems, efficient memory structures become crucial. The memory requirement can be expressed as:

$$M(n) = O\big(k \cdot h(n)\big),$$

where:

- n represents problem size;
- k is the tabu tenure;
- $h(n)$ is the memory needed per solution attribute.

2. Performance Optimization. The computational complexity typically follows:

$$T(n) = O\big(i \cdot n \cdot e(n)\big),$$

where:

- i is the number of iterations;
- $e(n)$ represents the cost of solution evaluation.

13.6.2   Parallel Implementations

Modern computing architectures enable parallel Tabu Search implementations through several strategies.

1. Parallel Neighborhood Exploration. Multiple threads can explore different parts of the neighborhood simultaneously:

$$N(x) = \bigcup_{p=1}^{P} N_p(x),$$

where:

- P is the number of parallel processes;
- $N_p(x)$ is the subset explored by process p.

2. Distributed Memory Management. Each processor maintains local memory structures:

$$M_p = T_p, F_p, E_p,$$

where:

- T_p represents local tabu lists;
- F_p stores frequency information;
- E_p maintains elite solutions.

13.6.3   Dynamic and Real-Time Applications

Modern applications often require handling dynamic conditions and real-time responses.

1. Dynamic Problem Adaptation. The objective function evolves over time:

$$f_t(x) = f(x,t),$$

requiring adaptive mechanisms:

- Dynamic tabu tenure adjustment;
- Real-time parameter tuning;
- Solution repair mechanisms.

2. Real-time Response Requirements. Time constraints impose additional considerations:

$$response\_time \leq \tau_{max},$$

necessitating:

- Early termination mechanisms;
- Quality-time trade-offs;
- Incremental solution updates.

13.6.4   Future Research Directions

The field continues to evolve with several promising research directions.

1. Integration with Machine Learning. Machine learning can enhance Tabu Search through:

$$prediction(x) = ML(history, features(x)),$$

where:

- *ML* represents a machine learning model;
- *history* contains past search experiences;
- *features(x)* extracts relevant solution characteristics.

2. Quantum Computing Applications. Quantum implementations may offer advantages in:

- Neighborhood exploration;
- Memory structure implementation;

- Parallel search processes.
- The potential speedup follows:

$$S_q(n) = O\left(\sqrt{N}\right),$$

- where N represents the size of the search space.

Hybrid Approaches: Modern trends favor hybrid methods combining Tabu Search with other techniques:

$$f'(x) = \alpha f_{TS}(x) + (1-\alpha) f_{other}(x),$$

where:

- f_{TS} represents the Tabu Search evaluation;
- f_{other} comes from complementary methods;
- α balances the contributions.

Key Research Challenges:

1. Theoretical Foundations:

   - Convergence properties in dynamic environments;
   - Performance bounds for parallel implementations;
   - Stability analysis of hybrid approaches.

2. Practical Advancements:

   - Automated parameter tuning mechanisms;
   - Robust handling of uncertainty;
   - Scalability to extremely large problems.

The future of Tabu Search lies in addressing these challenges while maintaining its core strengths of adaptiveness and effectiveness. Success in these areas will further establish Tabu Search as a fundamental tool in modern optimization.

This concluding section emphasizes both the practical aspects of implementing Tabu Search and its future potential. Understanding these considerations is crucial for practitioners seeking to apply Tabu Search effectively in real-world scenarios.

13.7   Practice and Exercises

13.7.1   Practice and Programming Exercises

All programming exercises for this chapter are available in the Google Colab notebook. You can access the complete notebook by scanning the QR code below or using this link: https://colab.research.google.com/drive/1hx1nQxP42SCIT2NPDU fg2cWGJlz2krxu

The notebook contains interactive Python code that you can run and modify to better understand Tabu Search concepts. Each exercise builds upon the previous ones, gradually introducing more complex aspects of the algorithm. Solutions to the exercises are available to instructors through the publisher's resources.

Exercise 1: Basic Tabu Search Implementation (Notebook Cells 1–3)

Implement Tabu Search for solving the Traveling Salesman Problem (TSP).

Tasks:
- Generate a random TSP instance with distance matrix;
- Implement basic Tabu Search components;
- Visualize the results;
- Analyze the optimized routes.

Learning Objectives:
- Understanding basic Tabu Search mechanisms;
- Implementing tabu list management;
- Developing neighborhood structures;
- Visualizing optimization results.

Exercise 2: Job Shop Scheduling with Tabu Search (Notebook Cells 4–7)

Apply Tabu Search to the Job Shop Scheduling problem.

Tasks:
- Implement solution representation for scheduling;
- Develop makespan calculation;
- Create appropriate neighborhood operators;
- Compare results with simple scheduling heuristics.

Learning Objectives:
- Understanding scheduling problem representation;
- Implementing complex objective functions;

- Developing problem-specific moves;
- Analyzing scheduling results.

Exercise 3: Advanced Features Implementation (Notebook Cells 8–10)

Enhance the basic Tabu Search with advanced features.

Tasks:
- Implement aspiration criteria;
- Add long-term memory structures;
- Develop intensification strategies;
- Create diversification mechanisms.

Learning Objectives:
- Understanding advanced Tabu Search concepts;
- Implementing memory structures;
- Developing search strategies;
- Analyzing improvement impact.

Note: Students should refer to the lecture materials and textbook sections covered in this chapter while working through these exercises. The accompanying Python notebook provides additional guidance and helper functions.

Exercise 4: Comparative Analysis (Notebook Cells 11–13)

Compare Tabu Search with other optimization methods.

Tasks:
- Implement multiple optimization methods;
- Compare performance metrics;
- Analyze convergence behavior;
- Visualize comparative results.

Programming Guidelines

When working with the exercises:

- Run the cells in order;
- Experiment with different parameters;
- Analyze the impact of your changes;
- Document your observations.

Additional Challenges

For each exercise, consider these extensions:

- Modify the tabu list structure;
- Implement different aspiration criteria;
- Add new neighborhood moves;
- Experiment with various problem instances.

Expected Outputs

For each exercise, you should:

- Produce visualization plots;
- Compare numerical results;
- Analyze convergence behavior;
- Document your findings.

13.7.2   Self-Assessment Questions

Instructions for Self-Assessment

Before attempting the questions:

- Review the theoretical foundations of Tabu Search;
- Complete all practical exercises;
- Understand both basic and advanced implementation aspects.

For each question:

- Read carefully and consider all options;
- Try to solve without referring to materials;
- Check your answers with the provided key;
- Review explanations for any incorrect answers.

Multiple Choice Questions

1. What is the primary purpose of the tabu list in Tabu Search?

    (a) To store all previously found solutions
    (b) To prevent cycling and promote exploration
    (c) To calculate the objective function value
    (d) To generate new candidate solutions

2. Which mechanism allows accepting a tabu move in certain circumstances?

   (a) Intensification strategy
   (b) Diversification mechanism
   (c) Aspiration criteria
   (d) Memory structure

3. In the context of Job Shop Scheduling, what does makespan represent?

   (a) Average processing time per job
   (b) Total completion time of all jobs
   (c) Maximum machine utilization
   (d) Minimum processing time

4. How does Advanced Tabu Search differ from Basic Tabu Search?

   (a) It only uses short-term memory
   (b) It runs fewer iterations
   (c) It combines multiple memory structures
   (d) It always finds the global optimum

5. What role does frequency-based memory play in Tabu Search?

   (a) Speeds up computation
   (b) Guides diversification
   (c) Reduces memory usage
   (d) Guarantees optimality

6. Which statement about move evaluation in Tabu Search is correct?

   (a) Only improving moves are accepted
   (b) Tabu moves are never accepted
   (c) Moves are evaluated based on multiple criteria
   (d) Random moves are preferred

7. What is a key advantage of Tabu Search over basic local search?

   (a) Guaranteed global optimum
   (b) Lower computational complexity
   (c) Ability to escape local optima
   (d) Simpler implementation

8. When implementing Tabu Search for TSP, what is a typical neighborhood move?

   (a) Adding new cities
   (b) 2-opt exchange
   (c) Random permutation
   (d) Path extension

9. How does solution diversity affect Tabu Search performance?

   (a) Higher diversity always leads to better solutions
   (b) Diversity balance helps exploration and exploitation

   (c) Lower diversity improves convergence speed
   (d) Diversity has no impact on solution quality

10. Which factor most significantly affects computation time in Tabu Search?

   (a) Tabu list size
   (b) Problem size and neighborhood structure
   (c) Initial solution quality
   (d) Number of local optima

Answers and Explanations

1. (b) The tabu list prevents revisiting recent solutions and forces exploration of new areas.
2. (c) Aspiration criteria allow accepting tabu moves if they lead to improved solutions.
3. (b) Makespan represents the total time needed to complete all operations.
4. (c) Advanced Tabu Search incorporates both short and long-term memory structures.
5. (b) Frequency-based memory helps guide the search toward unexplored regions.
6. (c) Moves are evaluated considering objective function, tabu status, and aspiration criteria.
7. (c) Tabu Search can escape local optima through its memory mechanisms.
8. (b) 2-opt exchange is a common neighborhood move for TSP optimization.
9. (b) Proper balance between diversity and intensity is crucial for performance.
10. (b) Neighborhood evaluation in larger problems dominates computational cost.

These questions cover the key concepts and practical aspects of Tabu Search implementation, emphasizing both theoretical understanding and practical application considerations.

13.8   Conclusion

Tabu Search provides a powerful and flexible optimization framework with continued relevance in modern computational challenges. Its strength comes from memory-based mechanisms that balance exploration and exploitation effectively. This balance enables efficient navigation of complex solution spaces while avoiding cycling and local optima traps.

Three key insights emerge from our examination of Tabu Search. First, adaptive memory structures form the foundation of the algorithm's effectiveness. Second, the method's flexibility allows customization to specific problem domains while maintaining core efficiency. Third, advanced features significantly enhance performance, though sometimes at increased computational cost.

Our practical exercises and case studies have demonstrated these principles in action. Applications to classical problems like the Traveling Salesman Problem and real-world challenges such as Job Shop Scheduling show how theoretical concepts translate into effective computational solutions.

References

1. Tsai, C.-W., Chiang, M.-C.: Chapter Six—Tabu search. In: Tsai, C.-W. and Chiang, M.-C. (eds.) Handbook of Metaheuristic Algorithms. pp. 95–109. Academic Press (2023). https://doi.org/10.1016/B978-0-44-319108-4.00019-8.
2. Glover, F.: Tabu Search—Part II. ORSA Journal on Computing. 2, 4–32 (1990). https://doi.org/10.1287/ijoc.2.1.4.
3. Glover, F.: Tabu Search—Part I. ORSA Journal on Computing. 1, 190–206 (1989). https://doi.org/10.1287/ijoc.1.3.190.
4. Glover, F.: Future paths for integer programming and links to artificial intelligence. Computers & Operations Research. 13, 533–549 (1986). https://doi.org/10.1016/0305-0548(86)90048-1.
5. Edelkamp, S., Schroedl, S.: Heuristic Search: Theory and Applications. Morgan Kaufmann, Amsterdam; Boston (2011).
6. Mariot, L., Leporati, A.: Heuristic Search by Particle Swarm Optimization of Boolean Functions for Cryptographic Applications. In: Proceedings of the Companion Publication of the 2015 Annual Conference on Genetic and Evolutionary Computation. pp. 1425–1426. Association for Computing Machinery, New York, NY, USA (2015). https://doi.org/10.1145/2739482.2764674.
7. Díaz, J.A., Fernández, E.: A Tabu search heuristic for the generalized assignment problem. European Journal of Operational Research. 132, 22–38 (2001). https://doi.org/10.1016/S0377-2217(00)00108-9.
8. Brandão, J.: A tabu search algorithm for the open vehicle routing problem. European Journal of Operational Research. 157, 552–564 (2004). https://doi.org/10.1016/S0377-2217(03)00238-8.

Chapter 14
Swarm Intelligence

Abstract This chapter explores Swarm Intelligence (SI), a computational paradigm drawing inspiration from nature's collective behaviors. SI algorithms simulate local interactions among simple agents to solve complex optimization problems. The material covers fundamental principles including stigmergy, self-organization, and feedback mechanisms. Core algorithms presented include Ant Colony Optimization, Particle Swarm Optimization, Artificial Bee Colony, Bacterial Foraging Optimization, and Firefly Algorithm. Each algorithm receives thorough examination through mathematical formulations, implementation guidelines, and practical coding examples. The chapter connects theoretical foundations with applications in diverse domains including routing, scheduling, pattern recognition, and industrial optimization. Through guided exercises and visualizations, readers gain both conceptual understanding and practical implementation skills.

14.1 Introduction to Swarm Intelligence

Swarm Intelligence (SI) represents a fascinating paradigm in artificial intelligence that draws inspiration from nature's collective behaviors [1, 2]. In the natural world, we observe remarkable examples of collective intelligence [3, 4]: ant colonies efficiently finding food sources, bird flocks maintaining complex flight formations, and bee colonies optimizing their resource gathering. These natural systems demonstrate how simple interactions between individuals can lead to sophisticated collective behavior.

14.1.1 Basic Principles and Natural Inspiration

The fundamental concept of Swarm Intelligence lies in the emergence of intelligent behavior through local interactions [5, 6]. Individual agents, following simple rules, create complex and adaptive collective behaviors without centralized control.

© The Author(s), under exclusive license to Springer Nature
Switzerland AG 2026

O. Kuznetsov, *Intelligent Systems: From Theory to Applications*, Cognitive
Technologies, https://doi.org/10.1007/978-3-032-00044-6_14

Consider an ant colony finding the shortest path to food. Each ant's simple behavior of following and depositing pheromones, when combined with others, creates an efficient path-finding system.

In mathematical terms, we can express a basic SI system as a tuple [2, 7]:

$$SI = (A, E, C, U),$$

where:

- $A = a_1, a_2, \ldots, a_n$ represents the set of agents;
- E denotes the environment in which agents operate;
- $C : A \times E \to A$ is the agent behavior function;
- $U : E \to \mathrm{R}$ is the utility or objective function.

14.1.2   Key Characteristics

- Decentralization forms the cornerstone of Swarm Intelligence. Unlike traditional control systems, no single agent directs the behavior of others. Each agent acts autonomously based on local information and simple rules. This decentralization provides robustness and adaptability to the system.
- Self-organization emerges naturally in SI systems. The system structure and functionality develop through internal dynamics rather than external control. Consider the mathematical representation of a self-organizing process:

$$x_{t+1} = f\left(x_t, N\left(x_t\right)\right),$$

- where x_t represents the system state at time t, and $N(x_t)$ denotes the local neighborhood interactions.

Emergence represents another crucial characteristic. Complex global behaviors arise from simple local interactions. This emergent behavior cannot be predicted by examining individual agents in isolation. The mathematical formulation of emergent behavior can be expressed as:

$$E : \prod_{i=1}^{n} B_i \to G,$$

where B_i represents individual behaviors and G denotes the global emergent behavior.

14.1.3   Mathematical Formalization

The dynamics of a swarm system typically follow this general form [2, 8]:

$$p_i(t+1) = p_i(t) + v_i(t) \quad v_i(t+1) = w \cdot v_i(t) + f\big(p_i(t), N_i(t)\big),$$

where:

- $p_i(t)$ represents the position of agent i at time t;
- $v_i(t)$ denotes the velocity or movement vector;
- w is an inertia weight;
- $f(p_i(t), N_i(t))$ captures the influence of local neighborhood interactions.

This formalization helps us understand how local interactions translate into global behavior patterns. Each agent updates its state based on its current condition and information from neighboring agents.

The collective behavior emerges through the interaction function [2, 8]:

$$I(t) = \frac{1}{n} \sum_{i=1}^{n} g_i\big(x_i(t)\big),$$

where $g_i(x_i(t))$ represents the contribution of agent i based on its local information x_i.

Through these mathematical foundations, we can analyze and design SI systems that effectively solve complex problems. The beauty of Swarm Intelligence lies in its ability to achieve sophisticated global behaviors through simple local rules, making it both powerful and practical for various optimization and control applications.

This introduction provides the groundwork for understanding more specific SI algorithms and their applications, which we will explore in subsequent sections. The principles outlined here form the basis for popular techniques such as Ant Colony Optimization and Particle Swarm Optimization, which we will examine in detail.

14.2   Theoretical Framework

The theoretical foundations of Swarm Intelligence rest on several key mechanisms that explain how individual agents interact to create collective intelligence. Understanding these mechanisms helps us design and implement effective swarm-based algorithms.

14.2.1   Stigmergy and Indirect Communication

Stigmergy represents a fundamental mechanism of coordination in swarm systems. First introduced by biologist Pierre-Paul Grassé in 1959 [9], stigmergy describes how agents communicate indirectly through modifications of their environment. This concept explains how simple agents can coordinate complex activities without direct communication [10].

We can formalize stigmergy as a function [10, 11]:

$$S : E \times A \to E,$$

where $S(e, a)$ represents the modified environment after agent a acts in environment e. For example, in ant colony optimization, the pheromone trail modification can be expressed as:

$$\tau_{ij}(t+1) = (1-\rho)\tau_{ij}(t) + \Delta\tau_{ij},$$

where τ_{ij} represents the pheromone level, ρ is the evaporation rate, and $\Delta\tau_{ij}$ is the amount of new pheromone deposited.

14.2.2 Positive and Negative Feedback Mechanisms

Feedback mechanisms play a crucial role in swarm behavior regulation. Positive feedback amplifies desirable actions, while negative feedback provides system stability.

The general feedback mechanism can be expressed as [10, 12]:

$$F(x) = \alpha x + \beta f(x),$$

where:

- $\alpha > 0$ represents positive feedback;
- $\beta < 0$ represents negative feedback;
- $f(x)$ is a nonlinear function describing agent interactions.

Positive feedback in ant foraging, for instance, strengthens successful paths [10, 12]:

$$\tau_{ij}^{+} = \tau_{ij} + q / L_{ij},$$

where q is the pheromone quantity and L_{ij} is the path length.

Negative feedback, such as pheromone evaporation, prevents premature convergence [10, 12]:

$$\tau_{ij}^{-} = (1-\rho)\tau_{ij}.$$

14.2.3 Exploration vs Exploitation Balance

The balance between exploration (searching new solutions) and exploitation (improving known solutions) is critical for swarm algorithm effectiveness. We can express this balance mathematically as [10, 12]:

$$P(a) = \lambda E(a) + (1-\lambda)S(a),$$

where:

- $P(a)$ is the probability of choosing action a;
- $E(a)$ is the exploration value;
- $S(a)$ is the exploitation value;
- $\lambda \in [0, 1]$ controls the balance.

A common implementation in PSO shows this balance [7, 8]:

$$v_i(t+1) = wv_i(t) + c_1 r_1 \left(p_i - x_i\right) + c_2 r_2 \left(g - x_i\right),$$

where w controls exploration through inertia, while c_1 and c_2 influence exploitation through personal and global best positions.

Collective Decision Making

Swarm systems make decisions through the aggregation of individual choices. The collective decision-making process can be modeled as [4, 8]:

$$D(t) = f\left(\frac{1}{n}\sum_{i=1}^{n} g_i\left(x_i(t)\right)\right),$$

where:

- $D(t)$ is the collective decision at time t;
- $g_i(x_i(t))$ represents individual agent contributions;
- f is an aggregation function.

In practice, this manifests as threshold-based decisions:

$$P(action) = \frac{T^n}{T^n + \theta^n},$$

where T is the stimulus level, θ is the response threshold, and n determines the steepness of the response curve.

The effectiveness of collective decision-making depends on:

- Information sharing quality;
- Individual decision accuracy;
- Group size;
- Diversity of options.

These theoretical frameworks provide the foundation for understanding how swarm intelligence emerges from simple interactions. They guide the design of practical algorithms and help explain why certain approaches work better than others in specific situations. The next sections will build upon these concepts to explore specific swarm intelligence algorithms and their applications.

14.3   Ant Colony Optimization (ACO)

Ant Colony Optimization represents one of the most successful and widely-applied swarm intelligence algorithms [1, 5]. This section explores its biological foundations, mathematical formulation, and practical implementation aspects.

14.3.1   Biological Inspiration and Principles

ACO draws inspiration from the foraging behavior of ant colonies in nature [1, 13]. Real ants find the shortest path between their nest and food sources through a process called stigmergy. When searching for food, ants deposit pheromones along their trail. Shorter paths accumulate more pheromone as ants traverse them more frequently. This simple mechanism leads to the emergence of optimal path finding.

The natural process translates into an optimization algorithm through three key principles:

- Indirect communication through pheromone trails;
- Preference for paths with higher pheromone concentrations;
- Positive feedback through pheromone accumulation.

14.3.2   Algorithm Formulation and Components

The ACO algorithm operates on a graph $G = (V, E)$, where V represents nodes and E represents edges. For each edge $(i, j) \in E$, we define [5, 14]:

- τ_{ij}: pheromone level;
- η_{ij}: heuristic information (typically $1/d_{ij}$, where d_{ij} is the distance).

The probability of an ant k moving from node i to node j is given by:

$$p_{ij}^k = \frac{\left(\tau_{ij}\right)^\alpha \left(\eta_{ij}\right)^\beta}{\sum_{l \in N_i^k} \left(\tau_{il}\right)^\alpha \left(\eta_{il}\right)^\beta},$$

where:

- N_i^k is the set of unvisited nodes accessible to ant k from node i;
- α and β control the relative importance of pheromone versus heuristic information.

14.3.3   *Pheromone Update Rules*

The pheromone update process consists of two steps:

1. Evaporation:

$$\tau_{ij} = \left(1 - \rho\right)\tau_{ij},$$

2. Deposit:

$$\tau_{ij} = \tau_{ij} + \sum_{k=1}^{m}\Delta\tau_{ij}^{k},$$

where:

- ρ is the evaporation rate ($0 < \rho < 1$);
- $\Delta\tau_{ij}^{k}$ is the amount of pheromone deposited by ant k.

14.3.4   *Parameter Selection and Tuning*

The effectiveness of ACO depends on careful parameter selection:

$$\alpha\left(\text{pheromone importance}\right): \text{Typically } 1-2$$

- Higher values increase exploitation of known good paths;
- Lower values promote exploration;

$$\beta\left(\text{heuristic importance}\right): \text{Typically } 2-5$$

- Higher values lead to more greedy search;
- Lower values allow more random exploration;

$$\rho\left(\text{evaporation rate}\right): \text{Typically } 0.1-0.3$$

- Higher values increase adaptation speed;
- Lower values promote stability;

$$Q\left(\text{pheromone deposit quantity}\right): \text{Problem} - \text{dependent}$$

- Affects the scale of pheromone updates;
- Should be proportional to solution quality.

Fig. 14.1 Ant Colony Optimization implementation strategies

14.3.5   Implementation Strategies

Here's a basic Python implementation framework for ACO:

```python
class ACO:
    def __init__(self, n_ants, n_iterations, alpha, beta, rho, Q):
        self.n_ants = n_ants
        self.n_iterations = n_iterations
        self.alpha = alpha
        self.beta = beta
        self.rho = rho
        self.Q = Q
        self.pheromone = None
        self.best_solution = None

    def solve(self, problem):
        self.initialize_pheromone(problem)
        for iteration in range(self.n_iterations):
            solutions = self.construct_solutions()
            self.update_pheromone(solutions)
            self.update_best_solution(solutions)
```

Figure 14.1 illustrates key strategies for implementing Ant Colony Optimization (ACO). It shows how pheromone trails guide ants to find optimal paths. The diagram highlights the balance between exploration and exploitation.

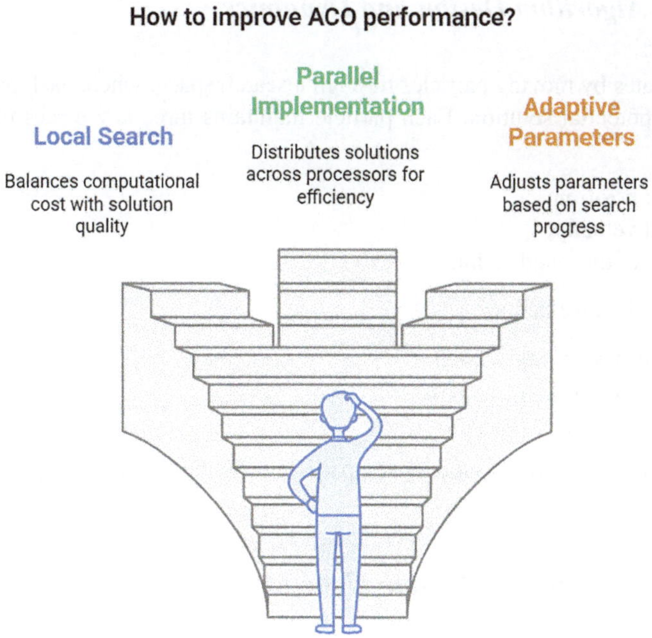

Fig. 14.2 Enhancing ACO performance

Figure 14.2 demonstrates techniques to improve ACO performance. It includes methods like parameter tuning and advanced update rules. These strategies help optimize the algorithm for specific problems.

ACO provides a powerful framework for solving complex optimization problems. Its success stems from effectively balancing exploration and exploitation through the pheromone mechanism. The algorithm's parameters offer flexibility in adapting to different problem characteristics, while various implementation strategies can enhance its performance for specific applications.

Understanding these components and their interactions allows practitioners to effectively implement and tune ACO algorithms for their specific optimization challenges. The next section will explore another major swarm intelligence algorithm: Particle Swarm Optimization.

14.4 Particle Swarm Optimization (PSO)

Particle Swarm Optimization represents another fundamental algorithm in swarm intelligence, inspired by the social behavior of bird flocks and fish schools [5, 15]. PSO combines individual learning with social interaction to search through complex solution spaces.

14.4.1   Algorithm Design and Dynamics

PSO operates by moving particles through a search space, where each particle represents a potential solution. Each particle maintains three key pieces of information [3, 4]:

- Current position;
- Current velocity;
- Best position found so far.

For a particle i, we define:

- Position: $x_i = (x_{i1}, x_{i2}, \ldots, x_{iD})$;
- Velocity: $v_i = (v_{i1}, v_{i2}, \ldots, v_{iD})$;
- Personal best: $p_i = (p_{i1}, p_{i2}, \ldots, p_{iD})$.

The global best position found by any particle is denoted as $g = (g_1, g_2, \ldots, g_D)$.

14.4.2   Velocity and Position Updates

The core PSO mechanism involves updating each particle's velocity and position. The standard velocity update equation is:

$$v_{id}(t+1) = w v_{id}(t) + c_1 r_1 \left(p_{id} - x_{id}(t) \right) + c_2 r_2 \left(g_d - x_{id}(t) \right),$$

where:

- w is the inertia weight;
- c_1 and c_2 are acceleration constants;
- r_1 and r_2 are random numbers in [0,1].

Position updates follow a simple rule:

$$x_{id}(t+1) = x_{id}(t) + v_{id}(t+1).$$

14.4.3   Parameter Settings and Variants

The performance of PSO depends heavily on parameter selection:
Inertia Weight (w):

- Typically in range [0.4, 0.9];
- Controls exploration vs exploitation;
- Can be dynamically adjusted:

$$w(t) = w_{\max} - \frac{w_{\max} - w_{\min}}{t_{\max}} t.$$

Acceleration Constants:

- c_1 (cognitive parameter): Usually 2.0;
- c_2 (social parameter): Usually 2.0;
- Balance personal and social learning.

Common PSO variants include:

- Constriction PSO:

$$\chi = \frac{2}{\left|2 - \phi - \sqrt{\phi^2 - 4\phi}\right|},$$

where $\phi = c_1 + c_2 > 4$;

- Adaptive PSO:

$$w(t) = w_{\min} + (w_{\max} - w_{\min})\exp(-\alpha t).$$

14.4.4 Handling Constraints

PSO must handle constraints effectively in practical applications. Common constraint handling approaches include:

1. Boundary Handling:

```
def handle_boundaries(position, v_max, bounds):
    for d in range(len(position)):
        if position[d] > bounds[d,1]:
            position[d] = bounds[d,1]
            velocity[d] *= -0.5
        elif position[d] < bounds[d,0]:
            position[d] = bounds[d,0]
            velocity[d] *= -0.5
```

2. Velocity Clamping:

$$v_{id} = sign(v_{id})\min\left(\|,v_{id}\|,\|,v_{\max}\right).$$

3. Penalty Functions:

$$f_{penalty}(x) = f(x) + \sum_{i=1}^{m} R_i \max\left(0, g_i(x)\right)^2.$$

14.4.5   Basic PSO Implementation

```python
class PSO:
    def __init__(self, n_particles, dimensions, bounds,
                 w=0.7, c1=2.0, c2=2.0):
        self.n_particles = n_particles
        self.dimensions = dimensions
        self.bounds = bounds
        self.w = w
        self.c1 = c1
        self.c2 = c2

        # Initialize particles
        self.positions = np.random.uniform(
            bounds[:,0], bounds[:,1],
            (n_particles, dimensions))
        self.velocities = np.zeros((n_particles, dimensions))
        self.personal_best = self.positions.copy()
        self.global_best = None

    def optimize(self, objective_function, max_iterations):
        for iteration in range(max_iterations):
            # Evaluate particles
            fitness = [objective_function(p) for p in self.
positions]

            # Update personal and global best
            self.update_best(fitness)

            # Update velocities and positions
            self.update_particles()

            # Handle constraints
            self.handle_constraints()
```

Figure 14.3 explains the movement of particles in PSO. It shows how each particle updates its position based on velocity and personal/global best positions. The illustration clarifies the core mechanics of PSO.

PSO's simplicity and effectiveness make it a popular choice for continuous optimization problems. Its ability to balance exploration and exploitation through simple update rules, combined with various enhancement strategies, provides a robust framework for solving complex optimization challenges. The next section will explore other swarm intelligence algorithms and their specific applications.

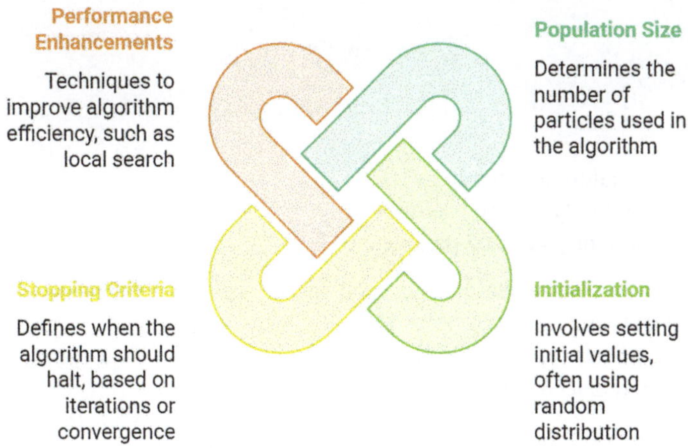

Fig. 14.3 Particle Swarm Optimization dynamics

14.5   Other Swarm Intelligence Algorithms

Beyond ACO and PSO, several other swarm intelligence algorithms offer unique approaches to optimization problems. Each algorithm draws inspiration from different natural phenomena and provides distinct advantages for specific types of problems.

14.5.1   Artificial Bee Colony (ABC)

The Artificial Bee Colony algorithm simulates the foraging behavior of honey bee colonies [13]. The algorithm divides artificial bees into three types: employed bees, onlooker bees, and scout bees [14].

Each food source position represents a possible solution. The nectar amount corresponds to the solution's quality. The probability of an onlooker bee choosing a food source is [16, 17]:

$$p_i = \frac{f\left(x_i\right)}{\sum_{j=1}^{SN} f\left(x_j\right)},$$

where:

- $f(x_i)$ is the fitness value of solution i;
- SN is the number of food sources.

The ABC algorithm follows these main phases:

1. Employed Bee Phase:

$$v_{ij} = x_{ij} + \phi_{ij}\left(x_{ij} - x_{kj}\right),$$

where:

- ϕ_{ij} is a random number in $[-1,1]$;
- k is a randomly selected solution;
- j is a randomly selected parameter.

2. Onlooker Bee Phase: Bees select food sources based on probability p_i and perform local search.
3. Scout Bee Phase: If a solution hasn't improved after a limit cycle, it is abandoned and replaced with a random solution:

$$x_{ij} = x_{\min,j} + rand\left(0,1\right)\left(x_{\max,j} - x_{\min,j}\right).$$

14.5.2   Bacterial Foraging Optimization (BFO)

BFO mimics the foraging strategy of E. coli bacteria [18]. The algorithm consists of four primary mechanisms [18, 19]:

1. Chemotaxis: Movement is described by:

$$\theta^i\left(j+1,k,l\right) = \theta^i\left(j,k,l\right) + C(i)\frac{\Delta(i)}{\sqrt{\Delta^T(i)\Delta(i)}},$$

where:

- θ^i represents the position of bacterium i;
- $C(i)$ is the step size;
- Δ is a random direction vector.

2. Swarming: Bacteria release attractants and repellents:

$$J_{cc}\left(\theta,P\left(j,k,l\right)\right) = \sum_{i=1}^{S} J_{cc}^i\left(\theta,\theta^i\left(j,k,l\right)\right).$$

3. Reproduction: Bacteria with poor performance die, while successful ones split.
4. Elimination-Dispersal: Some bacteria are eliminated and new ones are randomly initialized.

14.5.3 Firefly Algorithm (FA)

The Firefly Algorithm is inspired by the flashing behavior of fireflies [15, 20]. Each firefly's brightness relates to the objective function value, and attraction between fireflies depends on their relative brightness and distance [1, 15].

The attractiveness between fireflies is given by:

$$\beta(r) = \beta_0 e^{-\gamma r^2},$$

where:

- β_0 is the attractiveness at $r = 0$;
- γ is the light absorption coefficient;
- r is the distance between two fireflies.

The movement of firefly i attracted to brighter firefly j is:

$$x_i = x_i + \beta_0 e^{-\gamma r_{ij}^2}\left(x_j - x_i\right) + \alpha \epsilon_i,$$

where:

- α is the randomization parameter;
- ϵ_i is a vector of random numbers.

Implementation example for Firefly Algorithm:

```python
class FireflyAlgorithm:
    def __init__(self, n_fireflies, dimensions, bounds,
                 beta0=1.0, gamma=0.5, alpha=0.2):
        self.n_fireflies = n_fireflies
        self.dimensions = dimensions
        self.bounds = bounds
        self.beta0 = beta0
        self.gamma = gamma
        self.alpha = alpha

    def update_position(self, i, j):
        r = np.linalg.norm(self.positions[i] - self.positions[j])
        beta = self.beta0 * np.exp(-self.gamma * r**2)

        self.positions[i] += (beta * (self.positions[j] -
                              self.positions[i]) +
                              self.alpha * np.random.randn(self.
dimensions))
```

Figure 14.4 compares different swarm intelligence algorithms (ACO, PSO, ABC, BFO). It highlights their strengths and weaknesses. The comparison helps choose the right algorithm for specific tasks.

The choice of algorithm depends on specific problem characteristics:

- Problem dimensionality;
- Solution space characteristics;
- Computational resources available;
- Required solution accuracy.

Understanding these algorithms expands the toolset available for solving complex optimization problems. Each algorithm provides different mechanisms for balancing exploration and exploitation, making them suitable for different types of problems.

14.6   Practical Applications

Swarm Intelligence algorithms find extensive application across various domains [4, 13, 21]. This section explores key practical applications, emphasizing real-world implementations and their impact.

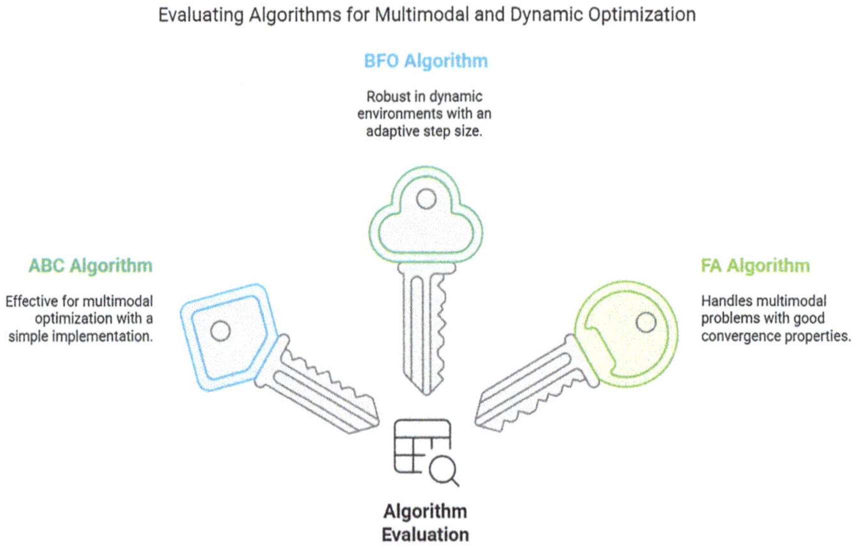

Fig. 14.4 Comparison of Swarm Intelligence Algorithms

14.6.1   Optimization Problems

Swarm Intelligence excels in solving complex optimization problems. In mathematical terms, these problems often take the form:

$$\min_{x \in S} f(x) \text{ subject to}: g_i(x) \le 0, i = 1,\ldots,m \, h_j(x) = 0, j = 1,\ldots,p.$$

Common applications include:

Function Optimization: The Rastrigin function demonstrates multimodal optimization: $f(x,y) = 20 + x^2 + y^2 - 10(\cos(2\pi x) + \cos(2\pi y))$

Example implementation for PSO:

```python
def optimize_rastrigin(dimensions=2, n_particles=30):
    bounds = [(-5.12, 5.12)] * dimensions
    pso = PSO(n_particles, dimensions, bounds)
    best_solution = pso.optimize(rastrigin_function)
    return best_solution
```

14.6.2   Routing and Scheduling

Transportation and logistics benefit significantly from swarm intelligence approaches. The Vehicle Routing Problem (VRP) represents a classic application:

$$\min \sum_{i=1}^{n} \sum_{j=1}^{n} c_{ij} x_{ij},$$

where:

- c_{ij} represents the cost between nodes i and j;
- x_{ij} is 1 if the route includes edge (i,j), 0 otherwise.

ACO implementation for routing:

```python
class ACORouter:
    def __init__(self, distances, n_ants=20):
        self.distances = distances
        self.n_ants = n_ants
        self.pheromone = np.ones_like(distances) / len(distances)

    def find_route(self, start, end):
        paths = self.construct_solutions(start, end)
        return self.select_best_path(paths)
```

14.6.3   Pattern Recognition

Swarm Intelligence algorithms effectively handle pattern recognition tasks. A common application is clustering:

$$\min \sum_{i=1}^{k} \sum_{x \in C_i} |x - \mu_i|^2,$$

where:

- C_i represents cluster I;
- μ_i is the centroid of cluster i.

PSO-based clustering example:

```python
def pso_clustering(data, n_clusters):
    dimensions = n_clusters * data.shape[1]
    pso = PSO(n_particles=20, dimensions=dimensions)
    centroids = pso.optimize(cluster_fitness)
    return assign_clusters(data, centroids)
```

14.6.4   Industrial Applications

Manufacturing and production systems benefit from swarm intelligence optimization. Consider a production scheduling problem:

$$\min \sum_{j=1}^{n} w_j T_j,$$

where:

- w_j is the priority of job j;
- T_j is the tardiness of job j.

Example applications include:

1. Load Balancing in Data Centers:

```python
def balance_load(servers, tasks):
    abc = ArtificialBeeColony(
        n_bees=50,
        n_food_sources=len(servers))
    return abc.optimize(load_distribution)
```

2. Energy Optimization in Smart Grids:

$$\min E_{total} = \sum_{t=1}^{T}\sum_{i=1}^{N} P_i\left(t\right),$$

where:

- $P_i(t)$ is the power consumption of unit i at time t.

3. Quality Control in Manufacturing

```python
def optimize_quality_parameters(process_data):
    fa = FireflyAlgorithm(
        n_fireflies=30,
        dimensions=len(process_parameters))
    return fa.optimize(quality_metric)
```

Implementation Considerations:

1. Real-time Constraints:

- Use simplified variants for time-critical applications;
- Implement parallel processing where possible;
- Consider hybrid approaches for faster convergence.

2. Scalability:

```python
def scale_solution(algorithm, problem_size):
    resources = estimate_resources(problem_size)
    return adapt_parameters(algorithm, resources)
```

3. Robustness:

- Implement error handling;
- Use adaptive parameters;
- Include fallback mechanisms.

Example of a robust industrial implementation:

```python
class IndustrialOptimizer:
    def __init__(self, algorithm_type, parameters):
        self.algorithm = self.initialize_algorithm(
            algorithm_type, parameters)
        self.backup_solution = None

    def optimize(self, problem):
        try:
            solution = self.algorithm.solve(problem)
```

```
        self.backup_solution = solution
        return solution
    except Exception as e:
        return self.handle_failure(e)
```

These practical applications demonstrate the versatility and effectiveness of swarm intelligence algorithms in solving real-world problems. The key to successful implementation lies in proper algorithm selection and parameter tuning based on specific problem characteristics and constraints.

14.7   Practice and Exercises

This section provides hands-on exercises to reinforce understanding of swarm intelligence algorithms through practical implementation and experimentation. All exercises are available in an interactive Python notebook that can be accessed using the QR code below or at: https://drive.google.com/file/d/1WlH6j3zEdecgKm6_g1RDb%2D%2DPMKY1tig5/view?usp=drive_link.

The notebook contains interactive Python code that you can run and modify to better understand swarm intelligence algorithms. Each exercise builds upon the previous ones, gradually introducing more complex aspects of algorithm implementation. Solutions to the exercises are available to instructors through the publisher's resources.

14.7.1   Practice and Programming Exercises

Exercise 1: Basic ACO Implementation

Implement a basic Ant Colony Optimization algorithm for path finding.

Tasks:
- Generate synthetic test data with known properties
- Implement core ACO algorithm components
- Visualize the optimization results
- Analyze the optimized parameters

Learning Objectives:
- Understanding basic ACO functionality
- Implementing parameter updates
- Visualizing optimization results

Exercise 2: Visual Analysis and Performance Tracking

Explore how ACO performs through detailed visualization.

Tasks:
- Implement enhanced visualization capabilities
- Track pheromone evolution
- Monitor solution diversity
- Study convergence behavior

Exercise 3: Parameter Analysis

Study how different parameters affect ACO performance.

Tasks:
- Implement parameter grid search
- Compare different parameter combinations
- Analyze convergence properties
- Visualize parameter interactions

Exercise 4: Optimization Algorithm Comparison

Compare ACO with PSO for optimization tasks.

Tasks:
- Implement PSO algorithm
- Compare convergence behavior
- Analyze solution quality
- Study performance characteristics

Exercise 5: Real-World Application

Apply ACO to a real-world routing problem.

Tasks:
- Handle real geographic data
- Create interactive visualizations
- Analyze route characteristics
- Study practical performance metrics

Programming Guidelines:
- Run cells in sequence
- Experiment with different parameters
- Analyze the impact of changes
- Document your observations

Additional Challenges:
- Try different city configurations
- Implement additional visualization methods
- Add new optimization metrics
- Compare with other algorithms

14.7.2   Self-Assessment Questions

Instructions

Before attempting these questions:

- Review the chapter material thoroughly.
- Complete all practical exercises.
- Understand both theoretical concepts and their practical applications.

Multiple Choice Questions

1. What is the primary mechanism behind Ant Colony Optimization (ACO)?

    (a) Direct communication between agents
    (b) Indirect communication through pheromone trails
    (c) Random exploration of search space
    (d) Centralized control by a leader agent

2. Which statement about Particle Swarm Optimization (PSO) is correct?

    (a) It always finds the global optimum regardless of parameter settings.
    (b) It balances exploration and exploitation using velocity updates.

    (c) It requires less memory than ACO.
    (d) It never explores more nodes than ACO.

3. What does stigmergy represent in swarm intelligence algorithms?

    (a) Direct communication between agents.
    (b) Indirect communication through environmental modifications.
    (c) The ability to find food sources quickly.
    (d) The use of centralized decision-making.

4. In Firefly Algorithm, what determines the attractiveness between two fireflies?

    (a) Their relative size and color.
    (b) Their distance and brightness.
    (c) The number of iterations completed.
    (d) The type of problem being solved.

5. Which characteristic defines swarm intelligence systems?

    (a) Centralized decision-making.
    (b) Emergence of global behavior from local interactions.
    (c) Single-agent problem-solving.
    (d) Use of gradient-based optimization.

6. What role does the inertia weight (w) play in PSO?

    (a) It controls the importance of heuristic information.
    (b) It balances exploration and exploitation.
    (c) It increases the speed of convergence.
    (d) It reduces computational overhead.

7. Which algorithm is best suited for solving clustering problems?

    (a) ACO
    (b) PSO
    (c) BFO
    (d) FA

8. What is the main advantage of using Artificial Bee Colony (ABC) over other algorithms?

    (a) Faster convergence in continuous optimization.
    (b) Better handling of discrete problems.
    (c) Simpler implementation compared to ACO.
    (d) Higher accuracy in pattern recognition.

9. What makes a search algorithm complete?

    (a) It always finds the optimal solution.
    (b) It uses minimal memory.
    (c) It guarantees finding a solution if one exists.
    (d) It explores all possible paths.

10. Which parameter in PSO controls the influence of social learning?

   (a) Inertia weight (w).
   (b) Cognitive parameter (c1).
   (c) Social parameter (c2).
   (d) Velocity clamping limit.

Answers

1. (b) ACO relies on indirect communication through pheromone trails. These trails guide ants toward promising paths. This is the key mechanism behind its success.
2. (b) PSO balances exploration and exploitation using velocity updates. The velocity equation incorporates both personal and global best positions. This ensures a balance between exploring new areas and refining known solutions.
3. (b) Stigmergy represents indirect communication through environmental modifications. For example, ants modify their environment by depositing pheromones, which others follow.
4. (b) In the Firefly Algorithm, attractiveness depends on distance and brightness. Brightness decreases with distance, guiding fireflies toward better solutions.
5. (b) Swarm intelligence systems are defined by the emergence of global behavior from local interactions. Individual agents follow simple rules, leading to complex collective behavior.
6. (b) The inertia weight (w) in PSO controls the balance between exploration and exploitation. Higher values promote exploration, while lower values encourage exploitation.
7. (b) PSO is well-suited for clustering problems because it efficiently searches for optimal cluster centroids. Its ability to handle continuous optimization makes it ideal for this task.
8. (a) ABC often converges faster in continuous optimization problems compared to other algorithms. This makes it particularly effective for certain types of optimization tasks.
9. (c) A search algorithm is complete if it guarantees finding a solution whenever one exists. Completeness is independent of optimality or memory usage.
10. (c) The social parameter (c2) in PSO controls the influence of social learning. It determines how much weight is given to the global best position when updating velocities.

14.8   Conclusions and Future Directions

Swarm Intelligence algorithms provide powerful tools for solving complex optimization problems. This chapter has shown how simple local interactions between agents produce sophisticated collective behaviors. The principles of decentralization, self-organization, and emergence enable these algorithms to navigate complex solution spaces effectively and adapt to diverse challenges.

Key contributions of swarm intelligence approaches include:

- First, these methods excel in balancing exploration and exploitation through natural mechanisms. Ant Colony Optimization uses pheromone trails to guide search while enabling exploration through randomness. Particle Swarm Optimization balances individual experience with social learning through its velocity update mechanism.
- Second, swarm algorithms demonstrate remarkable versatility across problem domains. Their applications span from discrete optimization in routing and scheduling to continuous parameter tuning in engineering design. The chapter has demonstrated their effectiveness in pattern recognition, industrial optimization, and resource allocation.
- Third, these bio-inspired approaches often succeed where traditional methods struggle. Their ability to escape local optima, handle constraints naturally, and adapt to changing environments makes them suitable for complex real-world problems.

Future research directions include developing automated parameter selection methods to reduce manual tuning requirements. Hybrid approaches combining swarm intelligence with machine learning techniques may enable more efficient search strategies. Adaptation for large-scale problems and enhanced real-time performance for dynamic environments represent important challenges.

References

1. Yang, X.-S., Karamanoglu, M.: Chapter 1—Nature-inspired computation and swarm intelligence: a state-of-the-art overview. In: Yang, X.-S. (ed.) Nature-Inspired Computation and Swarm Intelligence. pp. 3–18. Academic Press (2020). https://doi.org/10.1016/B978-0-12-819714-1.00010-5.
2. Kennedy, J., Eberhart, R.C., Shi, Y.: Chapter one—Models and Concepts of Life and Intelligence. In: Kennedy, J., Eberhart, R.C., and Shi, Y. (eds.) Swarm Intelligence. pp. 3–34. Morgan Kaufmann, San Francisco (2001). https://doi.org/10.1016/B978-155860595-4/50001-2.
3. Kennedy, J., Eberhart, R.C., Shi, Y.: Chapter four—Evolutionary Computation Theory and Paradigms. In: Kennedy, J., Eberhart, R.C., and Shi, Y. (eds.) Swarm Intelligence. pp. 133–185. Morgan Kaufmann, San Francisco (2001). https://doi.org/10.1016/B978-155860595-4/50004-8.
4. Sun, S., Liu, H.: 6—Particle Swarm Algorithm: Convergence and Applications. In: Yang, X.-S., Cui, Z., Xiao, R., Gandomi, A.H., and Karamanoglu, M. (eds.) Swarm Intelligence and Bio-Inspired Computation. pp. 137–168. Elsevier, Oxford (2013). https://doi.org/10.1016/B978-0-12-405163-8.00006-5.
5. Wang, C., Zhang, S., Ma, T., Xiao, Y., Chen, M.Z., Wang, L.: Swarm intelligence: A survey of model classification and applications. Chinese Journal of Aeronautics. 102982 (2024). https://doi.org/10.1016/j.cja.2024.03.019.

6. Krause, J., Ruxton, G.D., Krause, S.: Swarm intelligence in animals and humans. Trends in Ecology & Evolution. 25, 28–34 (2010). https://doi.org/10.1016/j.tree.2009.06.016.

7. Yang, X.-S., He, X.-S., Fan, Q.-W.: Chapter 7—Mathematical framework for algorithm analysis. In: Yang, X.-S. (ed.) Nature-Inspired Computation and Swarm Intelligence. pp. 89–108. Academic Press (2020). https://doi.org/10.1016/B978-0-12-819714-1.00017-8.

8. Yang, X.-S.: Chapter 5—Mathematical foundations for algorithm analysis. In: Yang, X.-S. (ed.) Nature-Inspired Computation and Swarm Intelligence. pp. 67–76. Academic Press (2020). https://doi.org/10.1016/B978-0-12-819714-1.00015-4.

9. Grassé, P.-P.: La reconstruction du nid et les coordinations interindividuelles chezBellicositermes natalensis etCubitermes sp. la théorie de la stigmergie: Essai d'interprétation du comportement des termites constructeurs. Ins. Soc. 6, 41–80 (1959). https://doi.org/10.1007/BF02223791.

10. Heylighen, F.: Stigmergy as a universal coordination mechanism I: Definition and components. Cognitive Systems Research. 38, 4–13 (2016). https://doi.org/10.1016/j.cogsys.2015.12.002.

11. Dipple, A., Raymond, K., Docherty, M.: General theory of stigmergy: Modelling stigma semantics. Cognitive Systems Research. 31–32, 61–92 (2014). https://doi.org/10.1016/j.cogsys.2014.02.002.

12. Nizamani, Q., Hashmani, A.A., Leghari, Z.H., Memon, Z.A., Munir, H.M., Novak, T., Jasinski, M.: Nature-inspired swarm intelligence algorithms for optimal distributed generation allocation: A comprehensive review for minimizing power losses in distribution networks. Alexandria Engineering Journal. 105, 692–723 (2024). https://doi.org/10.1016/j.aej.2024.08.033.

13. Rath, M., Darwish, A., Pati, B., Pattanayak, B.K., Panigrahi, C.R.: 2—Swarm intelligence as a solution for technological problems associated with Internet of Things. In: Hassanien, A.E. and Darwish, A. (eds.) Swarm Intelligence for Resource Management in Internet of Things. pp. 21–45. Academic Press (2020). https://doi.org/10.1016/B978-0-12-818287-1.00005-X.

14. Saka, M.P., Doğan, E., Aydogdu, I.: 2—Analysis of Swarm Intelligence–Based Algorithms for Constrained Optimization. In: Yang, X.-S., Cui, Z., Xiao, R., Gandomi, A.H., and Karamanoglu, M. (eds.) Swarm Intelligence and Bio-Inspired Computation. pp. 25–48. Elsevier, Oxford (2013). https://doi.org/10.1016/B978-0-12-405163-8.00002-8.

15. Yang, X.-S., Karamanoglu, M.: 1—Swarm Intelligence and Bio-Inspired Computation: An Overview. In: Yang, X.-S., Cui, Z., Xiao, R., Gandomi, A.H., and Karamanoglu, M. (eds.) Swarm Intelligence and Bio-Inspired Computation. pp. 3–23. Elsevier, Oxford (2013). https://doi.org/10.1016/B978-0-12-405163-8.00001-6.

16. Tong, M., Peng, Z., Wang, Q.: A hybrid artificial bee colony algorithm with high robustness for the multiple traveling salesman problem with multiple depots. Expert Systems with Applications. 260, 125446 (2025). https://doi.org/10.1016/j.eswa.2024.125446.

17. Liu, M., Yuan, Y., Xu, A., Deng, T., Jian, L.: A learning-based artificial bee colony algorithm for operation optimization in gas pipelines. Information Sciences. 690, 121593 (2025). https://doi.org/10.1016/j.ins.2024.121593.

18. Pang, S., Chen, M.-C.: Optimize railway crew scheduling by using modified bacterial foraging algorithm. Computers & Industrial Engineering. 180, 109218 (2023). https://doi.org/10.1016/j.cie.2023.109218.

19. Niu, B., Wang, H.: Bacterial Colony Optimization. Discrete Dynamics in Nature and Society. 2012, 698057 (2012). https://doi.org/10.1155/2012/698057.

20. Yang, X.-S., Zhao, Y.-X.: Chapter 3—Firefly algorithm and flower pollination algorithm. In: Yang, X.-S. (ed.) Nature-Inspired Computation and Swarm Intelligence. pp. 35–48. Academic Press (2020). https://doi.org/10.1016/B978-0-12-819714-1.00012-9.

21. Krause, J., Cordeiro, J., Parpinelli, R.S., Lopes, H.S.: 7—A Survey of Swarm Algorithms Applied to Discrete Optimization Problems. In: Yang, X.-S., Cui, Z., Xiao, R., Gandomi, A.H., and Karamanoglu, M. (eds.) Swarm Intelligence and Bio-Inspired Computation. pp. 169–191. Elsevier, Oxford (2013). https://doi.org/10.1016/B978-0-12-405163-8.00007-7.

Chapter 15
Introduction to Machine Learning

Abstract This chapter introduces the fundamentals of Machine Learning (ML), connecting theoretical principles with practical implementation. It covers supervised, unsupervised, and reinforcement learning paradigms with their mathematical foundations and algorithm designs. The material examines core concepts including bias-variance tradeoff, regularization, feature engineering, and model evaluation. Through Python implementations, readers learn to apply key algorithms such as k-Nearest Neighbors, linear regression, and clustering methods. The chapter addresses critical aspects of the ML workflow from data preparation to model deployment. Practical examples demonstrate solutions to real-world problems while highlighting performance optimization techniques. The content also examines ethical considerations including privacy protection and bias mitigation.

15.1 Introduction to Machine Learning

Machine Learning represents one of the most transformative technologies of our time [1, 2]. It forms a branch of artificial intelligence focused on developing algorithms that improve through experience. This introduction provides a foundation for understanding both the theoretical principles and practical applications of machine learning.

15.1.1 Historical Context and Evolution

The journey of machine learning began in the 1950s with Arthur Samuel's pioneering work on computer checkers [3]. Samuel defined machine learning as the field of study that gives computers the ability to learn without being explicitly programmed. This early definition remains remarkably relevant today [4, 5].

© The Author(s), under exclusive license to Springer Nature
Switzerland AG 2026
O. Kuznetsov, *Intelligent Systems: From Theory to Applications*, Cognitive
Technologies, https://doi.org/10.1007/978-3-032-00044-6_15

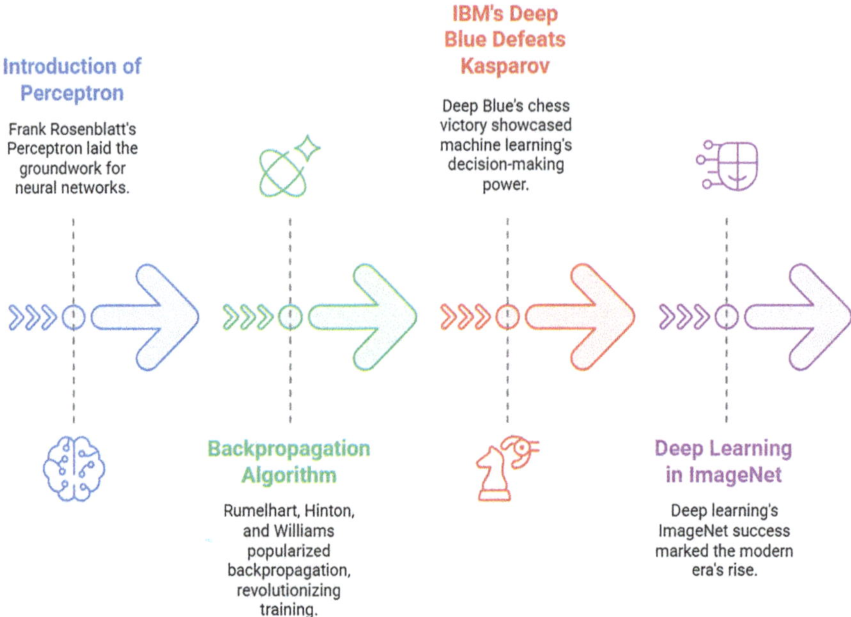

Fig. 15.1 Key milestones in machine learning history. This figure shows important events in the development of machine learning. It includes dates like 1957 when Perceptron was introduced, 1986 for backpropagation algorithm, 1997 for Deep Blue's victory, and 2012 for deep learning breakthrough

Several key milestones mark the evolution of machine learning (Fig. 15.1):

- 1957: Frank Rosenblatt introduced the Perceptron, establishing the foundation for modern neural networks.
- 1986: The publication of "Learning Internal Representations by Error Propagation" by Rumelhart, Hinton, and Williams popularized the backpropagation algorithm, revolutionizing neural network training.
- 1997: IBM's Deep Blue defeated world chess champion Garry Kasparov, demonstrating the potential of machine learning in complex decision-making tasks.
- 2012: Deep learning achieved breakthrough performance in the ImageNet competition, marking the beginning of the modern deep learning era.

15.1.2   Basic Definitions and Concepts

Machine learning focuses on building mathematical models that can learn patterns from data [6, 7]. The fundamental goal is to find a function $f : \mathcal{X} \to \mathcal{Y}$ that maps inputs to desired outputs.

For a given dataset

$$\mathcal{D} = \left(x_i, y_i \right)_{i=1}^{n},$$

where:

- $x_i \in \mathcal{X}$ represents input features;
- $y_i \in \mathcal{Y}$ represents target values;
- n is the number of training examples.

The learning process involves optimizing model parameters to minimize a loss function that measures prediction errors. A typical loss function for regression problems takes the form [8, 9]:

$$L\left(\theta \right) = \frac{1}{n} \sum_{i=1}^{n} \left(f_\theta \left(x_i \right) - y_i \right)^2,$$

where θ represents the model parameters.

15.1.3   Types of Machine Learning

Machine learning algorithms fall into three main categories (Fig. 15.2):

1. Supervised Learning: The algorithm learns from labeled data to predict outputs for new inputs. The training data consists of input-output pairs. Common applications include:

   - Classification (predicting discrete categories);
   - Regression (predicting continuous values).

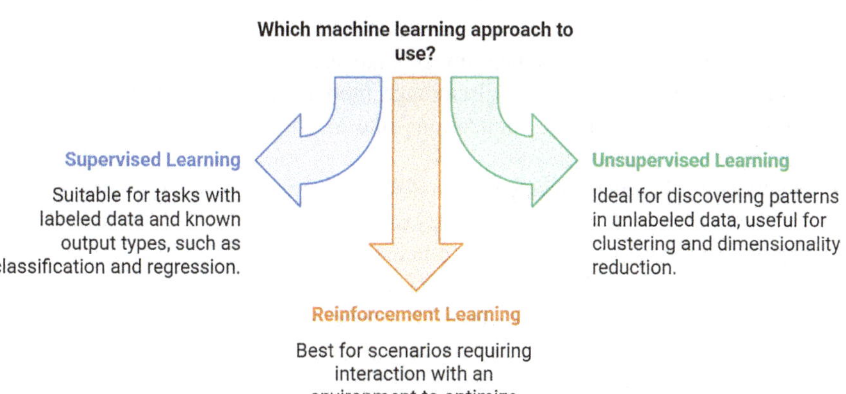

Fig. 15.2 Types of machine learning algorithms

2. Unsupervised Learning: The algorithm finds patterns in unlabeled data. It works
   with input data only, without explicit output values. Key applications include:

   - Clustering (grouping similar data points);
   - Dimensionality reduction (compressing data while preserving important
     information).

3. Reinforcement Learning: The algorithm learns through interaction with an environment. It receives rewards or penalties based on its actions and learns to optimize its behavior. This approach is formalized as a Markov Decision Process
   (MDP) [1, 2]:

$$(\mathcal{S},\mathcal{A},P,R,\gamma),$$

where:

   - $\mathcal{S}$ is the state space;
   - $\mathcal{A}$ is the action space;
   - P is the transition function;
   - R is the reward function;
   - γ is the discount factor.

Figure 15.2 illustrates three main categories of machine learning: supervised learning, unsupervised learning, and reinforcement learning. Each type is explained with
examples.

15.1.4 Current Applications and Impact

Machine learning has transformed numerous fields (Fig. 15.3):

- Computer Vision: Modern vision systems can recognize objects, faces, and
  actions in images and videos with remarkable accuracy. Applications range from
  autonomous vehicles to medical image analysis.
- Natural Language Processing: Language models can understand and generate
  human-like text, enabling applications like machine translation, chatbots, and
  document summarization.
- Healthcare: Machine learning aids in disease diagnosis, drug discovery, and personalized treatment planning. It analyzes medical images, patient records, and
  genetic data to improve healthcare outcomes.
- Financial Services: Algorithms detect fraudulent transactions, assess credit risk,
  and optimize investment portfolios. They process vast amounts of financial data
  in real-time to make informed decisions.
- Recommendation Systems: Online platforms use machine learning to personal-
  ize content and product recommendations, enhancing user experience and
  engagement.

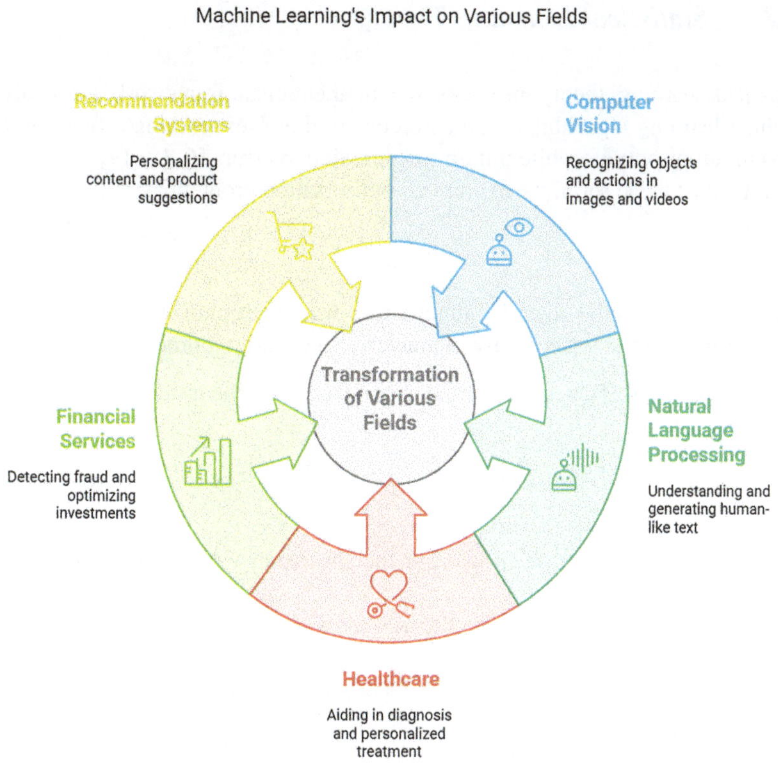

Fig. 15.3 Real-world applications of machine learning

Figure 15.3 demonstrates various fields where machine learning is applied successfully. Examples include computer vision, natural language processing, healthcare, finance, and recommendation systems.

These applications demonstrate how machine learning has become integral to modern technology. However, this impact also raises important ethical considerations regarding privacy, bias, and transparency. As we explore machine learning techniques in subsequent sections, we will consider both their technical aspects and broader societal implications.

15.2 Theoretical Foundations

The theoretical foundations of machine learning combine elements of statistics, information theory, optimization, and linear algebra [1, 2]. These foundations provide the framework for understanding how learning algorithms work and why they succeed or fail [10, 11].

15.2.1   Statistical Learning Theory

Statistical learning theory provides the mathematical framework for analyzing machine learning algorithms. The core concept involves learning a function f that maps inputs to outputs while minimizing prediction errors [6, 12, 13].

In its most basic form, we express an optimization problem as:

$$\min_{x \in \mathbb{R}^n} f(x),$$

where $f : \mathbb{R}^n \rightarrow \mathbb{R}$ is the objective function we want to minimize.

Two fundamental types of risk characterize the learning process:

1. Empirical Risk: This measures the average error on the training data:

$$R_{emp}(f) = \frac{1}{n} \sum_{i=1}^{n} L\big(f(x_i), y_i\big),$$

where L is a loss function measuring the discrepancy between predictions and actual values.

2. Expected Risk: This represents the true error over the entire data distribution:

$$R(f) = \mathbb{E}(x,y) \sim P(X,Y)\big[L\big(f(x),y\big)\big],$$

where $P(X, Y)$ is the joint probability distribution of the data.

The Principle of Empirical Risk Minimization (ERM) suggests finding:

$$f = \arg\min_{f \in \mathcal{F}} R_{emp}(f).$$

15.2.2   Bias-Variance Trade-Off

The bias-variance trade-off represents a fundamental concept in machine learning that describes the relationship between a model's complexity and its generalization ability. The total error of a model can be decomposed into three components [7, 14]:

$$Total\ Error = \big(Bias\big)^2 + Variance + Irreducible\ Error.$$

Bias represents the error from oversimplified assumptions in the learning algorithm. A high bias means the model misses relevant relations between features and target outputs (underfitting).

Variance represents the model's sensitivity to fluctuations in the training data. High variance indicates that the model captures random noise in the training data (overfitting).

The mathematical decomposition of the expected prediction error is:

$$\mathbb{E}\left[\left(y-\hat{f}(x)\right)^{2}\right] = \text{Bias}\left[\hat{f}(x)\right]^{2} + \text{Var}\left[\hat{f}(x)\right] + \sigma^{2},$$

where σ^{2} represents the irreducible error.

15.2.3   Feature Engineering and Selection

Feature engineering transforms raw data into format more suitable for machine learning algorithms [8, 9]. This process involves creating new features $\phi(x)$ from raw input data x.

Common feature engineering techniques include:

1. Standardization: Transform features to have zero mean and unit variance:

$$x' = \frac{x-\mu}{\sigma}.$$

2. Normalization: Scale features to a specific range (typically [0,1]):

$$x' = \frac{x-\min(x)}{\max(x)-\min(x)}.$$

Feature selection helps identify the most relevant features for model training. The objective function often includes a term for feature importance:

$$J(\theta) = L(\theta) + \lambda|\theta|_{p},$$

where $|\theta|_{p}$ represents the p-norm of the parameter vector.

15.2.4   Model Complexity and Regularization

Regularization helps control model complexity to prevent overfitting [5, 15]. It adds a penalty term to the loss function:

$$J(\theta) = L(\theta) + \lambda R(\theta),$$

where:

- $L(\theta)$ is the loss function;
- $R(\theta)$ is the regularization term;
- λ is the regularization strength.

Common regularization techniques include:

$$L1(\text{Lasso}): R(\theta) = |\theta|1 = \Sigma i = 1^n |\theta_i|$$

$$L2(\text{Ridge}): R(\theta) = |\theta|2^2 = \Sigma i = 1^n \theta_i^2$$

$$\text{Elastic Net}: R(\theta) = \alpha \sum_{i=1}^{n} |\theta_i| + (1-\alpha)\sum_{i=1}^{n}\theta_i^2$$

$$L1\ \text{Regularization}(\text{Lasso}): R(\theta) = |\theta|1 = \Sigma i = 1^n |\theta_i|;$$

$$L2\ \text{Regularization}(\text{Ridge}): R(\theta) = |\theta|2^2 = \Sigma i = 1^n \theta_i^2;$$

$$\text{Elastic Net combines both L1 and L2}: R(\theta) = \alpha |\theta|_1 + (1-\alpha)|\theta|_2^2;$$

where:

- n—total number of parameters;
- θ_i—i-th parameter of the model;
- $|\theta|_1$—L1 norm (sum of absolute values);
- $|\theta|_2^2$—square of L2-norm of parameter vector θ;
- $\sum_{i=1}^{n}\theta_i^2$—sum of $_{\text{squares}}$ of all model parameters;
- α—parameter controlling balance between L1 and L2 regularization ($0 \leq \alpha \leq 1$).

The choice of regularization method depends on the problem characteristics:

- L1 promotes sparsity (feature selection);
- L2 prevents large parameter values;
- Elastic Net combines both benefits.

Understanding these theoretical foundations is crucial for:

- Selecting appropriate algorithms for specific problems;
- Tuning model hyperparameters effectively;
- Diagnosing and addressing performance issues;
- Ensuring model generalization to new data.

These concepts form the basis for more advanced machine learning techniques we will explore in subsequent sections. They provide the tools needed to analyze and improve model performance in practical applications.

15.3   Main Types of Machine Learning

Machine learning algorithms can be categorized into three main paradigms based on how they learn from data. Each paradigm serves different purposes and is suited to specific types of problems.

15.3.1   Supervised Learning

Supervised learning works with labeled data to learn a mapping from inputs to outputs. In this approach, the algorithm learns from a training dataset $\mathcal{D} = \left(x_i, y_i \right)_{i=1}^{n}$, where x_i represents the input features and y_i represents the target values [8, 16].

Classification

In classification problems, the target variable y belongs to a discrete set of classes. A simple binary classification problem aims to find a function $f : \mathcal{X} \rightarrow 0,1$. The objective is to minimize the classification error:

$$\min_f \frac{1}{n} \sum_{i=1}^{n} L \left(f(x_i), y_i \right),$$

where L is often the cross-entropy loss:

$$L(y, \hat{y}) = -y \log(\hat{y}) - (1 - y) \log(1 - \hat{y}).$$

Common classification applications include:

- Email spam detection;
- Medical diagnosis;
- Image recognition;
- Sentiment analysis.

Regression

Regression problems involve predicting continuous values. The goal is to find a function $f : \mathcal{X} \rightarrow \mathbb{R}$. A typical regression problem minimizes the mean squared error:

$$\min_f \frac{1}{2n} \sum_{i=1}^{n} \left(f(x_i) - y_i \right)^2.$$

Linear regression represents the simplest form:

$$f(x) = \beta_0 + \beta_1 x_1 + \ldots + \beta_p x_p.$$

Common regression applications include:

- House price prediction;
- Stock market forecasting;

- Temperature prediction;
- Sales forecasting.

15.3.2   *Unsupervised Learning*

Unsupervised learning works with unlabeled data $\mathcal{D} = x_{i\,i=1}^{n}$. These algorithms discover patterns and structure within the data without explicit guidance [5, 16].

Clustering

Clustering algorithms group similar data points together. The k-means algorithm, for example, minimizes the within-cluster sum of squares:

$$\min_{S_1,\dots,S_K,\mu_1,\dots,\mu_K} \sum_{k=1}^{K}\sum_{x\in S_k} \left| x - \mu_k \right|^2 ,$$

where:

- S_k represents the k-th cluster;
- μ_k is the centroid of cluster k;
- K is the number of clusters.

Common clustering applications include:

- Customer segmentation;
- Document grouping;
- Image segmentation;
- Network analysis.

Dimensionality Reduction

These techniques transform high-dimensional data into a lower-dimensional representation while preserving important information. Principal Component Analysis (PCA) finds directions of maximum variance:

$$\max_{w:|w|=1} \mathrm{Var}\left(Xw\right).$$

Common dimensionality reduction applications include:

- Data visualization;
- Feature extraction;
- Noise reduction;

- Compression.

15.3.3   *Reinforcement Learning*

Reinforcement learning involves an agent learning to make decisions by interacting with an environment [5, 16]. The framework is formalized as a Markov Decision Process (MDP):

$$(\mathcal{S}, \mathcal{A}, P, R, \gamma).$$

Basic Concepts

The key elements include:

- State space $\mathcal{S}$: All possible situations;
- Action space $\mathcal{A}$: Available choices;
- Transition function P: Environment dynamics;
- Reward function R: Feedback signal;
- Discount factor γ: Future reward weighting.

The agent aims to maximize the expected cumulative reward:

$$\max_{\pi} \mathbb{E}\left[\sum_{t=0}^{\infty} \gamma^{t} R\left(s_{t}, a_{t}\right) \right],$$

where π is the policy mapping states to actions.

Applications

Reinforcement learning has found success in:

- Game playing (Chess, Go);
- Robotics control;
- Resource management;
- Autonomous driving;
- Portfolio optimization.

Practical Considerations

Each type of machine learning comes with its own challenges (Fig. 15.4).

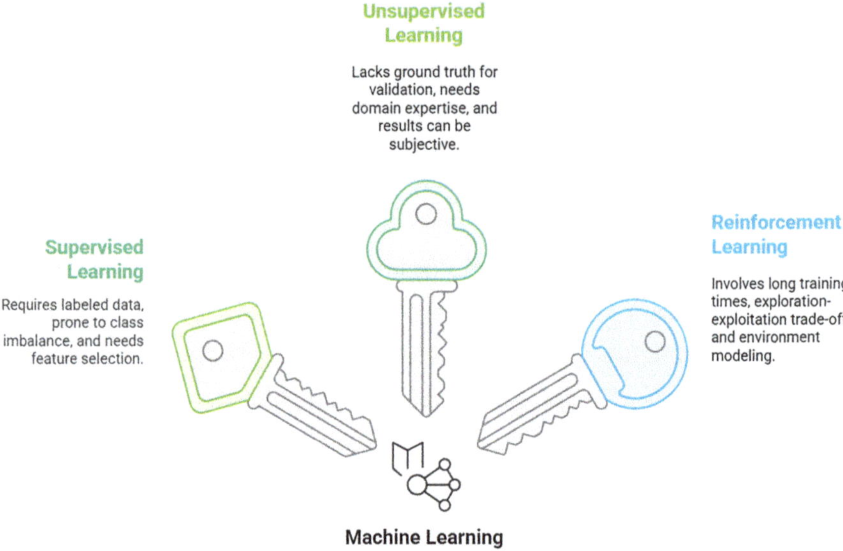

Fig. 15.4 Challenges in different learning paradigms

Figure 15.4 highlights common difficulties associated with each type of machine learning approach. It helps understand what factors to consider when choosing a method.

The choice of learning paradigm depends on:

- Available data type and quantity;
- Problem characteristics;
- Computational resources;
- Performance requirements;
- Interpretability needs.

Understanding these fundamental types of machine learning provides the foundation for selecting appropriate algorithms and designing effective solutions for specific problems.

15.4   The Machine Learning Process

The machine learning process follows a systematic approach from data preparation to model deployment [4, 5]. Understanding each step in this process is crucial for developing effective machine learning solutions.

15.4.1   *Data Preparation and Preprocessing*

Data preparation forms the foundation of successful machine learning applications. The quality and representation of input data significantly impact model performance.

Data Collection and Cleaning Raw data often contains missing values, outliers, and inconsistencies. Common cleaning steps include:

1. Missing Value Treatment:

   - Mean imputation:

$$x_{missing} = \frac{1}{n}\sum_{i=1}^{n} x_i;$$

   - Median imputation for skewed distributions;
   - Mode imputation for categorical variables.

2. Outlier Detection and Treatment:
   Using the Interquartile Range (IQR) method:

$$IQR = Q_3 - Q_1,$$

where:

   - Q_1 is the first quartile (25th percentile);
   - Q_3 is the third quartile (75th percentile).

   - Outliers are values of x for which:

$$x \langle Q_1 - 1.5 \cdot IQR \text{ or } x \rangle Q_3 + 1.5 \cdot IQR.$$

3. Feature Engineering:
   Creating new features $\phi(x)$ from raw data improves model performance:

   - Normalization:

$$x' = \frac{x - \min(x)}{\max(x) - \min(x)};$$

   - Standardization:

$$x' = \frac{x - \mu}{\sigma}.$$

4. Categorical Encoding:

   - One-hot encoding for nominal variables;
   - Label encoding for ordinal variables.

15.4.2   Model Selection and Training

Model selection involves choosing an appropriate algorithm based on the problem characteristics and data properties.

- Linear Models:

$$y = \beta_0 + \beta_1 x_1 + \ldots + \beta_p x_p;$$

- Neural Networks:

$$f(x) = \sigma\left(W_n \sigma\left(\ldots \sigma\left(W_1 x + b_1\right)\ldots\right) + b_n\right);$$

- Support Vector Machines:

$$f(x) = \sum_{i=1}^{n} \alpha_i y_i K\left(x_i, x\right) + b.$$

The training process minimizes a loss function $L(\theta)$ over the training data:

$$\min_\theta \frac{1}{n} \sum_{i=1}^{n} L\left(f_\theta\left(x_i\right), y_i\right) + \lambda R(\theta),$$

where $R(\theta)$ represents regularization.

15.4.3   Evaluation Metrics and Validation

Model evaluation requires appropriate metrics and validation strategies to assess performance.

Classification Metrics

$$\text{Accuracy : Accuracy} = \frac{\text{Correct Predictions}}{\text{Total Predictions}}.$$

Precision and Recall:

$$\text{Precision} = \frac{\text{True Positives}}{\text{True Positives} + \text{False Positives}}.$$

$$\text{Recall} = \frac{\text{True Positives}}{\text{True Positives} + \text{False Negatives}},$$

F1-Score:

$$F_1 = 2 \times \frac{\text{Precision} \times \text{Recall}}{\text{Precision} + \text{Recall}}.$$

Regression Metrics

Mean Squared Error (MSE):

$$\text{MSE} = \frac{1}{n} \sum_{i=1}^{n} \left(y_i - \hat{y}_i \right)^2.$$

R-squared:

$$R^2 = 1 - \frac{\sum_i \left(y_i - \hat{y}_i \right)^2}{\sum_i \left(y_i - \bar{y} \right)^2}.$$

Validation Techniques

K-fold Cross-validation divides data into k subsets:

- Split data into k folds;
- Train on $k - 1$ folds;
- Test on remaining fold;
- Repeat k times;
- Average results.

15.4.4   *Model Optimization and Tuning*

Model optimization involves adjusting hyperparameters to improve performance.

- Grid Search: Systematically works through multiple combinations of parameter tunes, cross-validating as it goes to determine which tune gives the best performance.
- Random Search: Samples parameter settings from specified distributions, often more efficient than grid search for high-dimensional spaces.
- Bayesian Optimization: Uses probabilistic models to guide the search for optimal hyperparameters:

$$x_{next} = \arg\max_x \text{acquisition}\big(\mu(x), \sigma(x)\big).$$

Practical Guidelines for Optimization

1. Learning Rate Selection: Start with a relatively large learning rate and decrease it when training stalls:

$$\eta_t = \frac{\eta_0}{1 + \text{decay}_r\text{ate} \times t}.$$

2. Batch Size Selection:

   - Smaller batches: Better generalization, slower training;
   - Larger batches: Faster training, potential generalization issues.

3. Early Stopping: Monitor validation performance and stop training when it starts to degrade:

   - Track validation error;
   - Set patience parameter;
   - Stop if no improvement for specified iterations.

Implementation Framework (Fig. 15.6)

1. Split data into training, validation, and test sets
2. Preprocess data appropriately
3. Select initial model architecture
4. Train model with cross-validation
5. Tune hyperparameters
6. Evaluate on test set
7. Deploy and monitor

Figure 15.5 presents a systematic approach to developing machine learning solutions. It covers data preparation, model selection, evaluation, optimization, and deployment stages.

Model optimization is an iterative process requiring careful balance between (Fig. 15.6).

Figure 15.6 shows elements that need balancing during model optimization. These include complexity, resources, performance requirements, and time constraints.

The success of machine learning projects depends heavily on following this systematic process while making appropriate adjustments based on specific problem requirements and constraints.

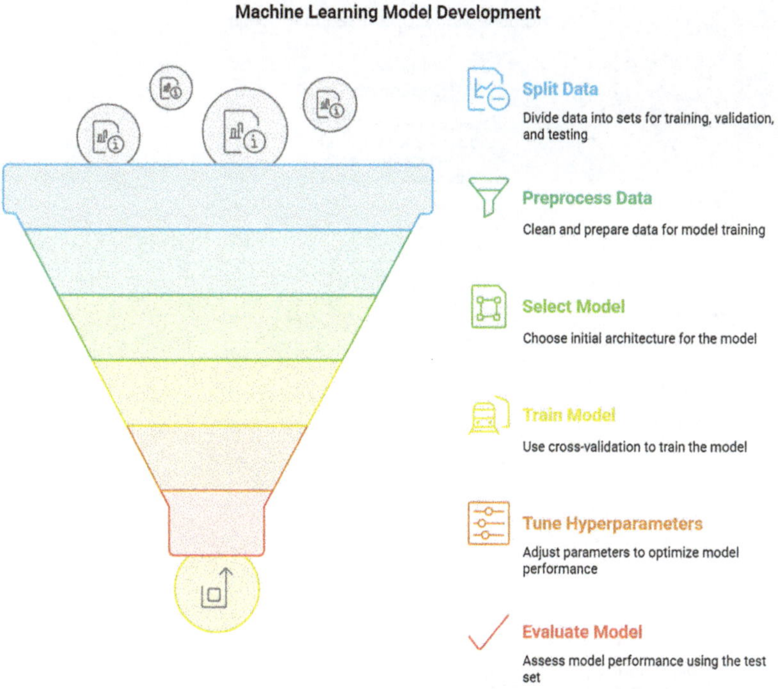

Fig. 15.5 Steps in the machine learning process

15.5   Practical Implementation

This section bridges theoretical concepts with practical implementation. We explore how to implement machine learning algorithms using Python, focusing on common challenges and their solutions.

15.5.1   Basic ML Algorithm Implementation (k-NN)

The k-Nearest Neighbors (k-NN) algorithm provides an excellent introduction to machine learning implementation. Its simplicity allows us to focus on fundamental concepts while building a working classifier.

Here is a basic implementation of k-NN:

```python
import numpy as np
from collections import Counter
class KNNClassifier:
    def __init__(self, k=3):
        self.k = k
```

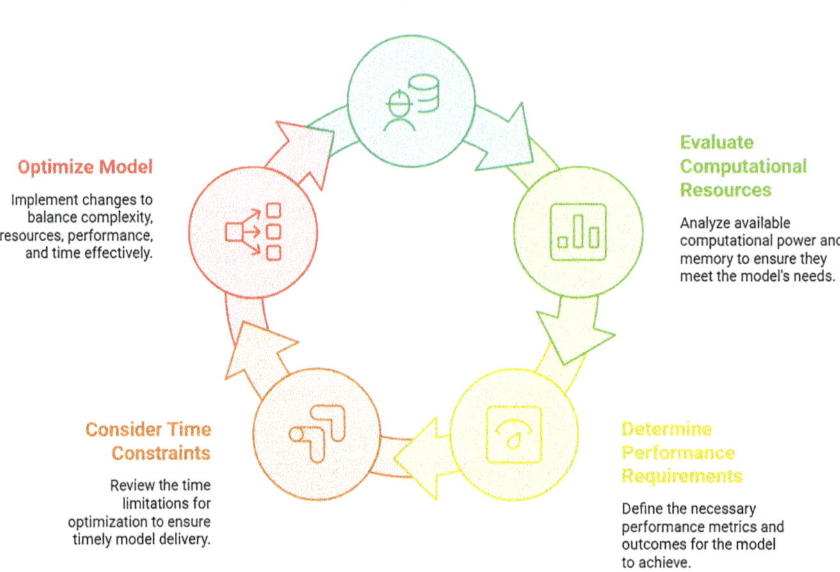

Fig. 15.6 Factors in model optimization

```python
def fit(self, X, y):
    """Store training data for distance calculations."""
    self.X_train = X
    self.y_train = y

def predict(self, X):
    """Predict class for each sample in X."""
    predictions = [self._predict(x) for x in X]
    return np.array(predictions)

def _predict(self, x):
    """Predict class for a single sample."""
    # Calculate distances to all training samples
    distances = [np.sqrt(np.sum((x - x_train)**2))
                for x_train in self.X_train]

    # Find indices of k nearest neighbors
    k_indices = np.argsort(distances)[:self.k]
```

```
    # Get their labels
    k_nearest_labels = [self.y_train[i] for i in k_indices]

    # Return most common class
    most_common = Counter(k_nearest_labels).most_common(1)
    return most_common[0][0]
```

The mathematics behind k-NN is straightforward. For a test point x, we:

- Calculate Euclidean distances:

$$d\left(x,x_i\right) = \sqrt{\sum_{j}\left(x_j - x_{ij}\right)^2};$$

- Find k points with minimum distances;
- Take majority vote of their labels.

15.5.2 Data Processing with Python

Effective data processing is crucial for model performance. Here's a systematic approach using Python:

```python
import pandas as pd
from sklearn.preprocessing import StandardScaler
from sklearn.model_selection import train_test_split
def process_data(data, target_column):
    """Process data for machine learning."""

    # Handle missing values
    data = data.fillna(data.mean())

    # Separate features and target
    X = data.drop(target_column, axis=1)
    y = data[target_column]

    # Scale features
    scaler = StandardScaler()
    X_scaled = scaler.fit_transform(X)

    # Split data
    X_train, X_test, y_train, y_test = train_test_split(
        X_scaled, y, test_size=0.2, random_state=42)

    return X_train, X_test, y_train, y_test, scaler
```

Key processing steps include:

- Missing value imputation;
- Feature scaling:

$$x' = \frac{x - \mu}{\sigma};$$

- Train-test splitting: Typically 80–20 split;
- Feature engineering when necessary.

15.5.3 *Model Evaluation in Practice*

Proper model evaluation requires both appropriate metrics and validation strategies:

```python
from sklearn.metrics import accuracy_score, precision_score,
recall_score
from sklearn.model_selection import cross_val_score
def evaluate_model(model, X, y):
    """Evaluate model using cross-validation."""

    # Perform 5-fold cross-validation
    cv_scores = cross_val_score(model, X, y, cv=5)

    print(f"Cross-validation scores: {cv_scores}")
    print(f"Mean CV score: {cv_scores.mean():.3f} "
          f"(+/- {cv_scores.std() * 2:.3f})")

    # Detailed metrics on test set
    y_pred = model.predict(X_test)

    accuracy = accuracy_score(y_test, y_pred)
    precision = precision_score(y_test, y_pred,
average='weighted')
    recall = recall_score(y_test, y_pred, average='weighted')

    return {
        'accuracy': accuracy,
        'precision': precision,
        'recall': recall
    }
```

Evaluation best practices:

- Use cross-validation to assess model stability;
- Calculate multiple metrics for comprehensive evaluation;
- Consider problem-specific metrics;
- Analyze errors patterns.

15.5.4   Common Pitfalls and Solutions

Several common issues arise in machine learning implementation (Fig. 15.7):

1. Data Leakage:

   - Problem: Test data influences training process;
   - Solution: Proper data splitting and careful feature engineering.

```python
# Wrong - leakage through scaling
scaler.fit_transform(X)  # Scales all data together
# Correct - separate scaling
scaler.fit(X_train)  # Fit on training data only
X_train_scaled = scaler.transform(X_train)
X_test_scaled = scaler.transform(X_test)
```

2. Overfitting:

   - Problem: Model performs well on training data but poorly on test data;
   - Solution: Cross-validation and regularization.

```python
from sklearn.linear_model import LogisticRegression
# Add regularization to prevent overfitting
model = LogisticRegression(C=1.0)  # C is inverse of
regularization strength
```

3. Class Imbalance:

   - Problem: Uneven class distribution affects model performance;
   - Solution: Resampling or weighted loss functions.

```python
from sklearn.utils.class_weight import compute_class_weight
# Compute balanced weights
weights = compute_class_weight('balanced',
                               classes=np.unique(y),
                               y=y)
# Use weights in model training
model = LogisticRegression(class_weight='balanced')
```

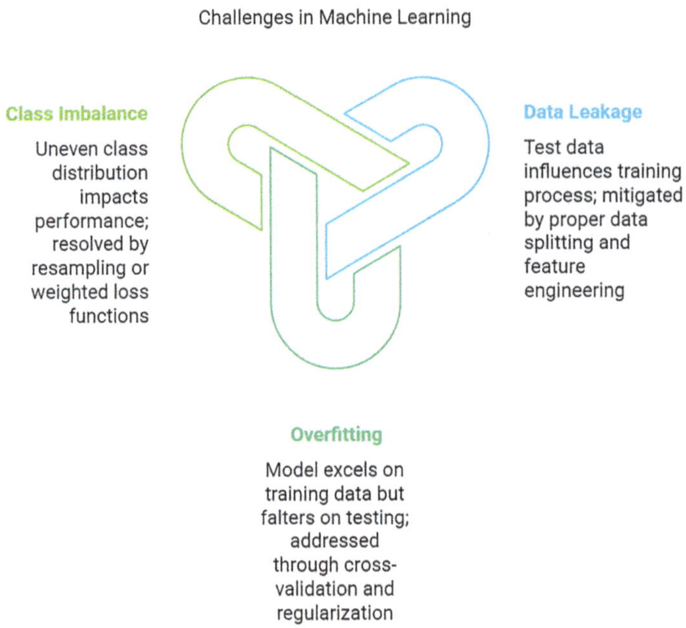

Fig. 15.7 Common pitfalls in machine learning implementation

Figure 15.7 points out frequent problems in practical implementation. Solutions for issues like data leakage, overfitting, and class imbalance are provided.

Figure 15.8 summarizes guidelines for building robust models. Recommendations cover starting simple, organizing code well, optimizing performance, and thorough testing.

By following these practices and being aware of common pitfalls, you can develop more robust and reliable machine learning solutions.

## 15.6	Case Study: Building a Recommendation System

This case study demonstrates the complete machine learning process through the practical example of building a recommendation system. We will explore both technical implementation and ethical considerations.

### 15.6.1	Problem Definition

Recommendation systems suggest items to users based on their preferences and behavior. Our goal is to build a system that predicts user ratings for items they haven't yet rated.

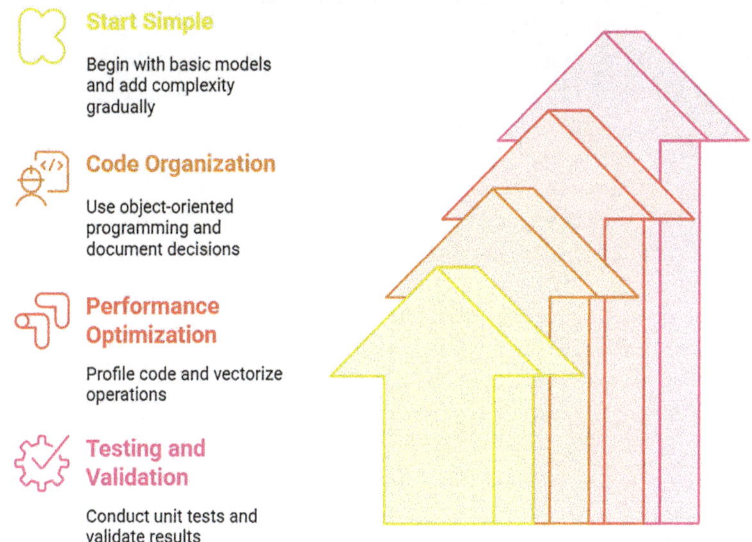

Fig. 15.8 Best practices for reliable machine learning systems

Formally, we aim to learn a function:

$$f : \mathcal{U} \times \mathcal{I} \to \mathbb{R},$$

where:

- $\mathcal{U}$ is the set of users;
- I is the set of items;
- The output represents predicted ratings.

The data typically takes the form of a sparse matrix R, where r_{ui} represents the rating of user u for item i.

15.6.2   Data Processing

Let's implement data processing for a movie recommendation system:

```python
import pandas as pd
import numpy as np
from sklearn.model_selection import train_test_split
def process_rating_data(ratings_df):
    """Process rating data for recommendation system."""
```

```python
    # Create user and item mappings
    user_ids = ratings_df['user_id'].unique()
    item_ids = ratings_df['item_id'].unique()

    user_to_idx = {id: idx for idx, id in enumerate(user_ids)}
    item_to_idx = {id: idx for idx, id in enumerate(item_ids)}

    # Convert to matrix format
    n_users = len(user_ids)
    n_items = len(item_ids)
    ratings_matrix = np.zeros((n_users, n_items))

    for _, row in ratings_df.iterrows():
        user_idx = user_to_idx[row['user_id']]
        item_idx = item_to_idx[row['item_id']]
        ratings_matrix[user_idx, item_idx] = row['rating']

    return ratings_matrix, user_to_idx, item_to_idx
```

Key processing steps include:

- Handling missing values;
- Creating user and item indices;
- Normalizing ratings;
- Splitting into training and validation sets.

15.6.3   *Model Implementation*

We implement a matrix factorization model using Singular Value
Decomposition (SVD):

```python
class MatrixFactorization:
    def __init__(self, n_factors=50, learning_rate=0.01,
regularization=0.02):
        self.n_factors = n_factors
        self.learning_rate = learning_rate
        self.regularization = regularization

    def fit(self, R, n_epochs=20):
        n_users, n_items = R.shape

        # Initialize latent factors

self.P = np.random.normal(0, 0.1, (n_users, self.n_factors))
```

```python
self.Q = np.random.normal(0, 0.1, (n_items, self.n_factors))

        # Optimize using SGD
        for epoch in range(n_epochs):
            for u in range(n_users):
                for i in range(n_items):
                    if R[u, i] > 0:  # Only train on
observed ratings
                        error = R[u, i] - self.predict(u, i)

                        # Update latent factors
                        self.P[u] += self.learning_rate * (
                            error * self.Q[i] - self.
regularization * self.P[u])
                        self.Q[i] += self.learning_rate * (
                            error * self.P[u] - self.
regularization * self.Q[i])

    def predict(self, user_idx, item_idx):
        return np.dot(self.P[user_idx], self.Q[item_idx])
```

The mathematical model is based on matrix factorization:

$$R \oplus PQ^{T},$$

where:

- $P \in \mathbb{R}^{n_{users} \times k}$ represents user latent factors;
- $Q \in \mathbb{R}^{n_{items} \times k}$ represents item latent factors;
- k is the number of latent factors.

15.6.4   Performance Evaluation

We evaluate the recommendation system using multiple metrics:

```python
def evaluate_recommendations(model, R_test, R_train=None):
    """Evaluate recommendation system performance."""

    # Calculate RMSE
    n_users, n_items = R_test.shape
    squared_error = 0
    n_ratings = 0

    for u in range(n_users):
```

```
        for i in range(n_items):
            if R_test[u, i] > 0:
                pred = model.predict(u, i)
                squared_error += (R_test[u, i] - pred) ** 2
                n_ratings += 1

    rmse = np.sqrt(squared_error / n_ratings)


    # Calculate coverage
    n_possible = n_users * n_items
    n_predictions = np.sum(R_test > 0)
    coverage = n_predictions / n_possible

    return {
        'rmse': rmse,
        'coverage': coverage
    }
```

Key evaluation metrics include:

- Root Mean Square Error (RMSE):

$$RMSE = \sqrt{\frac{1}{|T|} \sum_{(u,i) \in T} \left(r_{ui} - \hat{r}_{ui} \right)^2 } \ ;$$

- Coverage: Proportion of user-item pairs that can be predicted;
- Novelty: Ability to recommend non-obvious items;
- Diversity: Variety in recommendations.

15.6.5   Ethical Considerations

Recommendation systems raise important ethical concerns (Fig. 15.9).

Figure 15.9 represents potential ethical challenges in recommendation systems. Aspects like privacy protection and bias mitigation are discussed.

Privacy Protection

- Minimize personal data collection;
- Implement data anonymization;
- Provide user control over data.

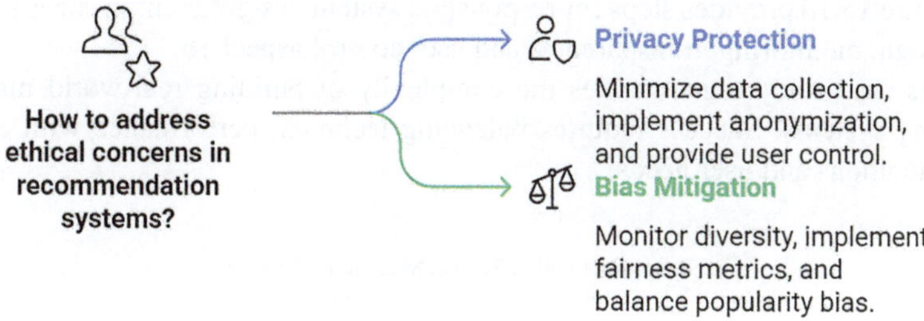

Fig. 15.9 Ethical concerns in recommendation systems

```python
def anonymize_user_data(ratings_df):
    """Anonymize user data while preserving rating patterns."""

    # Create anonymous user IDs
    user_mapping = {old_id: f"user_{idx}"
                    for idx, old_id in enumerate(ratings_df['user_
id'].unique())}

    # Apply mapping
    anon_df = ratings_df.copy()
    anon_df['user_id'] = anon_df['user_id'].map(user_mapping)

    return anon_df
```

Bias Mitigation

- Monitor recommendation diversity;
- Implement fairness metrics;
- Balance popularity bias.

```python
def calculate_diversity_metrics(recommendations):
    """Calculate diversity metrics for recommendations."""

    unique_items = len(set(recommendations))

category_distribution = pd.Series(recommendations).value_
counts(normalize=True)
    entropy = -sum(p * np.log(p) for p in category_distribution)

    return {
        'unique_items': unique_items,
        'entropy': entropy
    }
```

Figure 15.10 provides steps for responsible system design. It emphasizes privacy by design, monitoring, transparency, and user control aspects.

This case study demonstrates the complexity of building real-world machine learning systems. Success requires balancing technical performance with ethical considerations and user needs.

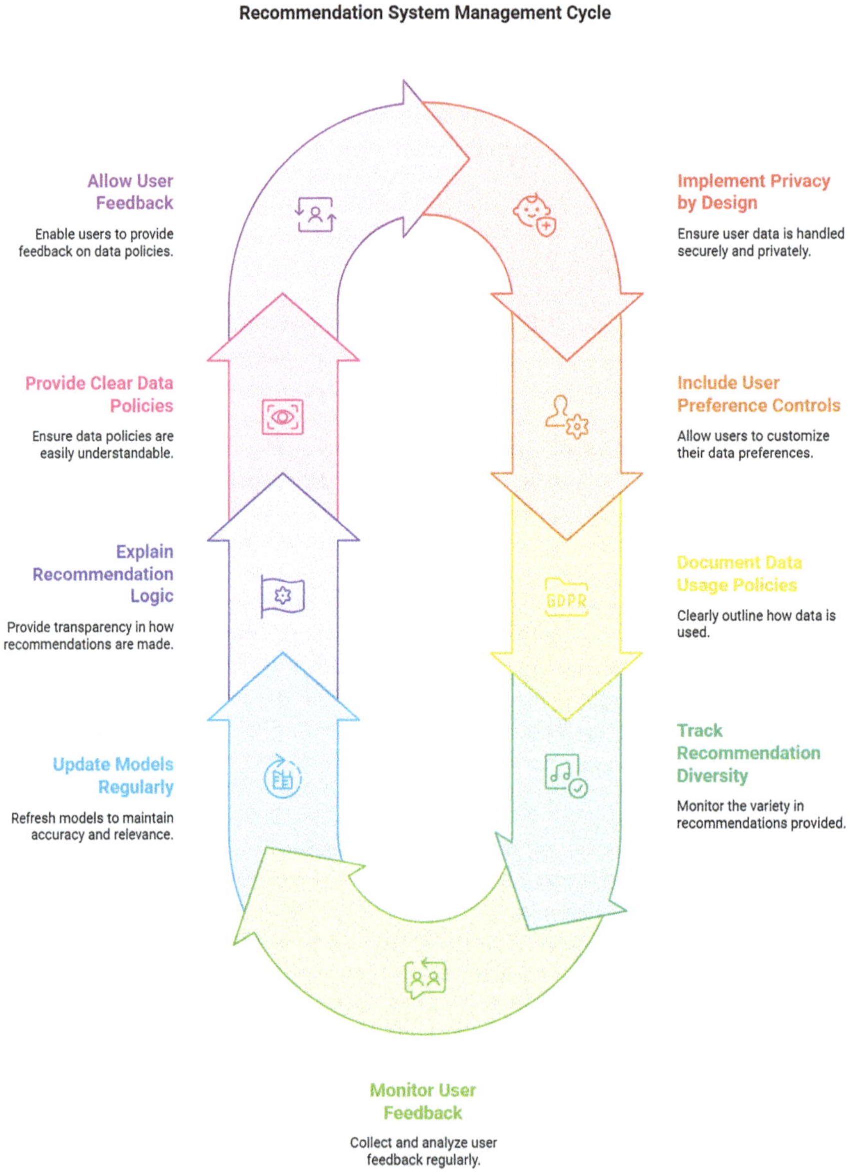

Fig. 15.10 Guidelines for ethical machine learning systems

15.7 Practice and Exercises

This section provides hands-on exercises to reinforce understanding of machine learning concepts through practical implementation. The exercises build upon the theoretical foundations presented earlier in the chapter. Each exercise focuses on specific aspects of machine learning. They help develop practical skills.

All exercises are available in an interactive Python notebook that can be accessed using the QR code below or at: https://colab.research.google.com/drive/1s4kFNCoGDuMNMhjfwO340wD_MuXYkOl8

The accompanying Python notebook contains interactive code examples. These can be run and modified. This helps better understand machine learning algorithms. All exercises are designed to gradually introduce more complex topics. Solutions to the exercises are available to instructors through the publisher's resources.

15.7.1 Programming Exercises

Exercise 1: Basic Machine Learning Implementation

Learning Objectives:
- Understand the core components of supervised learning algorithms.
- Implement a basic linear regression model with visualization.
- Learn to handle synthetic data and evaluate model performance using Mean Squared Error (MSE).

Tasks:

1. Generate synthetic data for a simple linear relationship with noise.
2. Split the data into training and testing sets.
3. Train a linear regression model on the training set.
4. Evaluate the model's performance using MSE.
5. Visualize the true values and predictions on a scatter plot.

Exercise 2: Classification Algorithm Development

Learning Objectives:
- Master the implementation of the k-Nearest Neighbors (k-NN) algorithm.
- Understand how to calculate distances and classify samples based on majority voting.
- Visualize decision boundaries and interpret results using a confusion matrix.

Tasks:

1. Generate a synthetic classification dataset with two features and two classes.
2. Implement the k-NN classifier from scratch.
3. Train the model and make predictions on the test set.
4. Evaluate accuracy and visualize decision boundaries.
5. Analyze the confusion matrix to assess performance.

Exercise 3: Unsupervised Learning Techniques

Learning Objectives:
- Understand clustering techniques such as k-Means and dimensionality reduction methods like PCA.
- Learn to evaluate clustering performance using Silhouette Score.
- Visualize clusters and reduced-dimensional data effectively.

Tasks:

1. Generate a synthetic dataset with four clusters.
2. Apply k-Means clustering and evaluate performance using Silhouette Score.
3. Perform PCA to reduce dimensions and visualize the transformed data.
4. Reapply k-Means on PCA-transformed data and compare results.

Exercise 4: Model Evaluation and Optimization

Learning Objectives:
- Learn to optimize hyperparameters using Grid Search and cross-validation.
- Compare baseline and optimized models using accuracy and other metrics.
- Interpret results and identify the best parameters for improved performance.

Tasks:

1. Generate a synthetic dataset for binary classification.
2. Train a baseline logistic regression model and evaluate its accuracy.
3. Define a hyperparameter grid and perform Grid Search with cross-validation.
4. Train an optimized model using the best parameters and evaluate its performance.
5. Create a bar chart comparing baseline and optimized accuracies.

Exercise 5: Real-World Application Case Study

Learning Objectives:
- Apply machine learning techniques to solve a real-world problem (sentiment analysis).
- Preprocess text data using TF-IDF vectorization.
- Train a logistic regression model and interpret results using feature importance analysis.
- Address ethical considerations such as anonymization and bias mitigation.

Tasks:

1. Load the IMDB movie reviews dataset and preprocess the text data.
2. Convert text data into numerical features using TF-IDF vectorization.
3. Train a logistic regression model and evaluate its performance using accuracy, precision, recall, and F1-score.
4. Identify top positive and negative features influencing sentiment predictions.
5. Anonymize sensitive information in reviews to ensure privacy protection.

15.7.2   Self-Assessment Questions

Instructions

Before attempting these questions, review the chapter material thoroughly, complete all practical exercises, and understand both theoretical concepts and their implementations. Take time to answer each question before checking the solutions.

Multiple Choice Questions

1. What is the main goal of supervised learning?

   (a) To group similar data points together.
   (b) To predict outputs for new inputs based on labeled training data.
   (c) To optimize rewards in an environment.
   (d) To reduce the dimensionality of data.

   Correct Answer: (b) To predict outputs for new inputs based on labeled training data.

2. Which algorithm is best suited for clustering unlabeled data into distinct groups?

   (a) Linear Regression
   (b) k-Means Clustering
   (c) Logistic Regression
   (d) Reinforcement Learning

Correct Answer: (b) k-Means Clustering

3. What does the Silhouette Score measure in clustering algorithms?

   (a) The distance between centroids.
   (b) The quality of clustering by balancing cohesion and separation.
   (c) The accuracy of predictions.
   (d) The explained variance of principal components.

Correct Answer: (b) The quality of clustering by balancing cohesion and separation.

4. In reinforcement learning, what does the reward function represent?

   (a) The cost associated with incorrect predictions.
   (b) Feedback from the environment indicating success or failure.
   (c) The probability distribution of the data.
   (d) The regularization strength of a model.

Correct Answer: (b) Feedback from the environment indicating success or failure.

5. What is the purpose of feature scaling in machine learning?

   (a) To increase the number of features.
   (b) To ensure all features contribute equally to the model.
   (c) To anonymize sensitive data.
   (d) To reduce the size of the dataset.

Correct Answer: (b) To ensure all features contribute equally to the model.

6. Which metric is commonly used to evaluate regression models?

   (a) Accuracy
   (b) Mean Squared Error (MSE)
   (c) Precision
   (d) Recall

Correct Answer: (b) Mean Squared Error (MSE)

7. What is the role of hyperparameter tuning in machine learning?

   (a) To preprocess the data.
   (b) To optimize model performance by adjusting parameters.
   (c) To visualize decision boundaries.
   (d) To generate synthetic datasets.

Correct Answer: (b) To optimize model performance by adjusting parameters.

8. Which regularization technique promotes sparsity in linear models?

   (a) L1 Regularization (Lasso)
   (b) L2 Regularization (Ridge)
   (c) Elastic Net
   (d) Dropout

Correct Answer: (a) L1 Regularization (Lasso)

9. What is the primary challenge in unsupervised learning?

   (a) Lack of labeled data.
   (b) High computational complexity.
   (c) Overfitting to training data.
   (d) Ethical considerations.

Correct Answer: (a) Lack of labeled data.

10. Which component of reinforcement learning determines the next action to take?

   (a) State Space
   (b) Action Space
   (c) Policy
   (d) Reward Function

Correct Answer: (c) Policy

15.8 Conclusions and Future Directions

Machine learning provides powerful tools for solving complex problems across diverse domains. This chapter has demonstrated how these algorithms excel in analyzing patterns, making predictions, and optimizing decisions through both theoretical foundations and practical implementations.

Three key insights emerge from our examination of machine learning. First, supervised learning excels in classification and regression tasks by learning from labeled data. Unsupervised learning discovers hidden structures through clustering and dimensionality reduction. Reinforcement learning enables agents to learn optimal behaviors through environmental interaction.

Second, practical implementation requires careful attention to data preparation, feature engineering, model selection, and evaluation strategies. Effective preprocessing significantly impacts model performance. Proper validation ensures generalization to new data. Visualization tools aid in interpreting results and identifying improvements.

Third, ethical considerations must guide machine learning development. Privacy protection, bias mitigation, and transparency are crucial when deploying these systems. Responsible AI development requires balancing technical performance with societal values and user needs.

Looking forward, several research directions show promise. Advanced algorithms like deep learning architectures and ensemble methods continue expanding capabilities. Hybrid approaches combining machine learning with other AI techniques offer new solutions for complex problems. Applications in autonomous systems, personalized medicine, climate modeling, and smart cities highlight machine learning's importance in addressing global challenges. As datasets grow, efficient algorithms and specialized hardware will play vital roles in maintaining performance.

References

1. Norvig, P.: Artificial intelligence: a modern approach. Global edition. Pearson, Boston (2021).
2. Russell, S., Norvig, P.: Artificial Intelligence: A Modern Approach. Pearson, Hoboken (2020).
3. Samuel, A.L.: Some Studies in Machine Learning Using the Game of Checkers. IBM Journal of Research and Development. 3, 210–229 (1959). https://doi.org/10.1147/rd.33.0210.
4. Sterling, T., Anderson, M., Brodowicz, M.: Chapter 19—Machine Learning. In: Sterling, T., Anderson, M., and Brodowicz, M. (eds.) High Performance Computing (Second Edition). pp. 383–393. Morgan Kaufmann, Boston/Waltham (MA) (2025). https://doi.org/10.1016/B978-0-12-823035-0.00019-5.
5. Liu, Z., Gu, Z., Liu, P.: Chapter 5—Machine learning basics. In: Liu, Z., Gu, Z., and Liu, P. (eds.) Transportation Big Data. pp. 129–175. Elsevier (2025). https://doi.org/10.1016/B978-0-443-33891-5.00005-3.
6. Natarajan, B.K.: 8—Generalizing the Learning Model. In: Natarajan, B.K. (ed.) Machine Learning. pp. 167–196. Morgan Kaufmann, San Francisco (CA) (1991). https://doi.org/10.1016/B978-0-08-051053-8.50011-9.
7. Theodoridis, S.: Chapter 3—Learning in Parametric Modeling: Basic Concepts and Directions. In: Theodoridis, S. (ed.) Machine Learning. pp. 53–103. Academic Press, Oxford (2015). https://doi.org/10.1016/B978-0-12-801522-3.00003-3.
8. Vieira, S., Lopez Pinaya, W.H., Mechelli, A.: Chapter 2—Main concepts in machine learning. In: Mechelli, A. and Vieira, S. (eds.) Machine Learning. pp. 21–44. Academic Press (2020). https://doi.org/10.1016/B978-0-12-815739-8.00002-X.
9. Cohen, S.: Chapter 2—The basics of machine learning: Strategies and techniques. In: Chauhan, C. and Cohen, S. (eds.) Artificial Intelligence in Pathology (Second Edition). pp. 15–42. Elsevier (2025). https://doi.org/10.1016/B978-0-323-95359-7.00002-9.
10. Gori, M.: Chapter 6—Learning and Reasoning With Constraints. In: Gori, M. (ed.) Machine Learning. pp. 340–444. Morgan Kaufmann (2018). https://doi.org/10.1016/B978-0-08-100659-7.00006-3.
11. Maas, A.L., Daly, R.E., Pham, P.T., Huang, D., Ng, A.Y., Potts, C.: Learning Word Vectors for Sentiment Analysis. In: Lin, D., Matsumoto, Y., and Mihalcea, R. (eds.) Proceedings of the 49th Annual Meeting of the Association for Computational Linguistics: Human Language Technologies. pp. 142–150. Association for Computational Linguistics, Portland, Oregon, USA (2011).
12. Natarajan, B.K.: 5—Learning Functions. In: Natarajan, B.K. (ed.) Machine Learning. pp. 99–124. Morgan Kaufmann, San Francisco (CA) (1991). https://doi.org/10.1016/B978-0-08-051053-8.50008-9.
13. Natarajan, B.K.: 4—Learning Concepts on Uncountable Domains. In: Natarajan, B.K. (ed.) Machine Learning. pp. 73–97. Morgan Kaufmann, San Francisco (CA) (1991). https://doi.org/10.1016/B978-0-08-051053-8.50007-7.
14. Natarajan, B.K.: 2—Learning Concepts on Countable Domains. In: Natarajan, B.K. (ed.) Machine Learning. pp. 7–40. Morgan Kaufmann, San Francisco (CA) (1991). https://doi.org/10.1016/B978-0-08-051053-8.50005-3.
15. Gori, M.: Chapter 2—Learning Principles. In: Gori, M. (ed.) Machine Learning. pp. 60–121. Morgan Kaufmann (2018). https://doi.org/10.1016/B978-0-08-100659-7.00002-6.
16. Vieira, S., Lopez Pinaya, W.H., Mechelli, A.: Chapter 1—Introduction to machine learning. In: Mechelli, A. and Vieira, S. (eds.) Machine Learning. pp. 1–20. Academic Press (2020). https://doi.org/10.1016/B978-0-12-815739-8.00001-8.

Chapter 16
Supervised Learning

Abstract This chapter examines supervised learning, a core machine learning paradigm where models learn from labeled examples to make predictions on new data. It covers the complete supervised learning workflow from data preparation to model deployment. The material first establishes mathematical foundations including probability theory, linear algebra, and optimization principles essential for understanding learning algorithms. Classification algorithms such as logistic regression, support vector machines, decision trees, and random forests are presented with their mathematical formulations and Python implementations. Regression techniques for predicting continuous values are explored through linear, polynomial, and regularized models. The chapter provides systematic approaches to model evaluation using appropriate metrics and validation strategies for ensuring generalization. Through hands-on exercises and real-world examples, readers develop practical skills in implementing supervised learning solutions while understanding their theoretical underpinnings.

16.1 Foundations of Supervised Learning

Supervised learning represents a fundamental paradigm in machine learning where a model learns from labeled examples to make predictions on new data [3, 4]. This section introduces the core concepts and framework of supervised learning, establishing a foundation for more advanced topics.

16.1.1 Introduction to Supervised Learning

Supervised learning is a machine learning approach in which a model learns a mapping between inputs and desired outputs based on example pairs [5, 6]. Consider a dataset

O. Kuznetsov, *Intelligent Systems: From Theory to Applications*, Cognitive Technologies, https://doi.org/10.1007/978-3-032-00044-6_16

$$D = (x_1, y_1), (x_2, y_2), \ldots, (x_n, y_n),$$

where $x_i \in X$ are input vectors and $y_i \in Y$ are corresponding target values. The goal is to learn a function $f: X \to Y$ that can predict outputs for new, unseen inputs.

The term "supervised" comes from the idea that the learning process is guided by the correct answers (labels) provided in the training data. This is analogous to a student learning with the help of a teacher who provides examples and correct solutions.

16.1.2   The Supervised Learning Framework

The supervised learning process follows a structured approach [7, 8]:

1. Data Collection: Gather labeled examples that represent the problem space.
2. Data Preprocessing: Clean and prepare the data for model training. This includes:

   - Handling missing values
   - Normalizing features
   - Encoding categorical variables

3. Model Selection: Choose an appropriate algorithm based on the problem type and data characteristics.
4. Training: The model learns patterns from the training data by minimizing a loss function $L(f(x), y)$.
5. Validation: Evaluate model performance on unseen data to ensure generalization.
6. Deployment: Apply the trained model to make predictions on new data.

16.1.3   Types of Supervised Learning Problems

Supervised learning problems fall into two main categories [4, 6]:

Classification

In classification problems, the goal is to assign input examples to discrete categories. The target variable y belongs to a finite set of classes: $Y = 1, 2, \ldots, K$. Examples include:

- Email spam detection (binary classification);
- Image recognition (multi-class classification);
- Disease diagnosis (multi-label classification).

Regression

Regression problems involve predicting continuous numerical values. Here, $Y \subseteq R$. Common applications include [3, 4]:

- House price prediction;
- Temperature forecasting;
- Sales volume estimation.

16.1.4   Components of a Supervised Learning System

A supervised learning system consists of several essential components (Fig. 16.1):

1. Feature Space (X):

   - Input variables that describe each example;
   - Can include numerical or categorical data;
   - Often requires preprocessing and normalization;

2. Target Space (Y):

   - Output variables to be predicted;
   - Discrete for classification;
   - Continuous for regression;

3. Training Dataset:

   - Labeled examples used for model learning;
   - Must be representative of the problem space;
   - Usually split into training and validation sets;

4. Model (f):

   - Mathematical function that maps inputs to outputs;
   - Contains learnable parameters adjusted during training;
   - Chosen based on problem requirements;

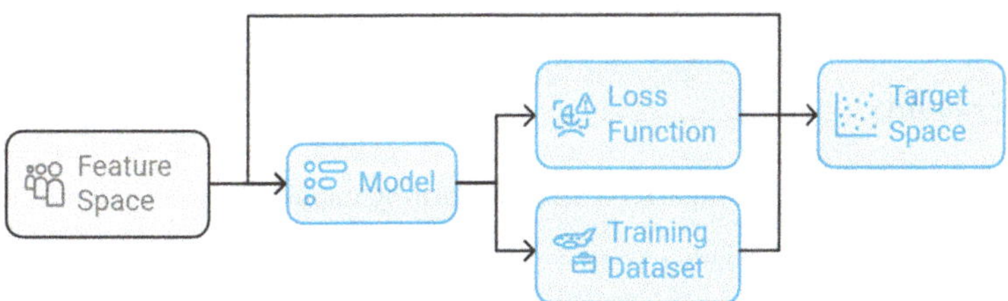

Fig. 16.1 Components of a supervised learning system

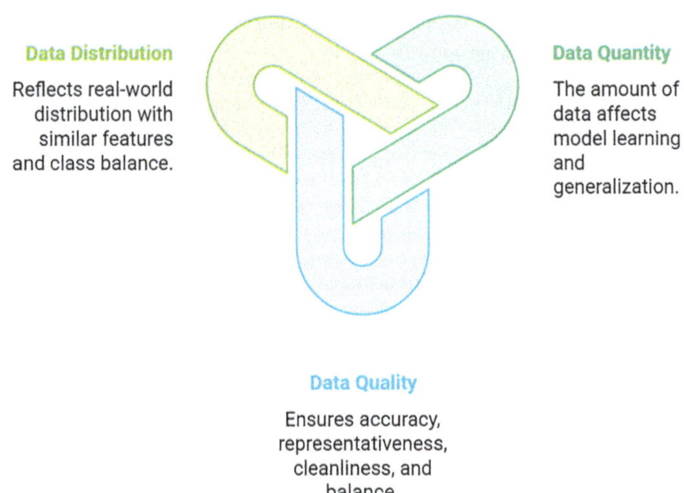

Fig. 16.2   The role of training data in model performance

5. Loss Function (L):

- Measures prediction errors;
- Guides the learning process;
- Examples: Mean Squared Error, Cross-Entropy.

Figure 16.1 illustrates the key elements of a supervised learning model. It shows how input features, target outputs, a learning function, and an evaluation metric interact.

16.1.5   The Role of Training Data

Training data quality significantly impacts model performance. Key considerations include (Fig. 16.2):

Figure 16.2 explains how data quantity, quality, and distribution impact model accuracy. High-quality, well-balanced datasets lead to better generalization.

Data Quantity

The amount of training data affects model learning capability:

- More data generally leads to better generalization;
- Required quantity depends on problem complexity;
- Must balance against computational constraints.

Data Quality

High-quality training data should be:

- Accurate: Correctly labeled examples;
- Representative: Covers the full problem space;
- Clean: Free from noise and errors;
- Balanced: Equal representation of classes.

Data Distribution

Training data should reflect the real-world distribution:

- Similar feature distributions;
- Appropriate class balance;
- Realistic noise levels.

The relationship between training examples can be formally expressed as:

$$P(x,y) = P(y|x)P(x),$$

where $P(x, y)$ represents the joint probability distribution of inputs and outputs.

Supervised learning provides a powerful framework for developing predictive models. Understanding its fundamental components and requirements is essential for successful implementation. The next sections will build upon these foundations to explore specific algorithms and practical applications in detail.

16.2 Mathematical Foundations

Supervised learning builds upon several fundamental mathematical concepts [5, 6]. This section explores the essential mathematical tools needed to understand and implement supervised learning algorithms effectively.

16.2.1 Probability Theory and Statistics

Probability theory provides the framework for modeling uncertainty in supervised learning [1, 2].

Conditional Probability

The probability of an event given another event forms the basis of many supervised learning models:

$$P(Y|X) = \frac{P(X,Y)}{P(X)}.$$

This fundamental relationship describes how we predict outputs (Y) based on inputs (X). In supervised learning, we estimate this conditional probability from training data.

Bayes' Theorem

Bayes' theorem connects prior knowledge with observed data:

$$P(Y|X) = \frac{P(X|Y)P(Y)}{P(X)},$$

This theorem helps us understand how to update our predictions based on new evidence. Many classification algorithms implicitly use this principle.

Probability Distributions

Common distributions in supervised learning include:

1. Gaussian Distribution:

$$f(x) = \frac{1}{\sigma\sqrt{2\pi}} e^{-\frac{(x-\mu)^2}{2\sigma^2}}.$$

2. Bernoulli Distribution:

$$P(X = k) = p^k (1-p)^{1-k}, k \in 0,1.$$

These distributions often model feature values and classification outcomes.

16.2.2   Linear Algebra Essentials

Linear algebra provides tools for handling high-dimensional data and implementing learning algorithms.

Vectors and Matrices

Data in supervised learning is typically represented as vectors and matrices:

- Input features: $x = [x_1, x_2, \ldots, x_n]^T$;

- Weight matrix: $W = [w_{ij}]_{m \times n}$;
- Output vectors: $y = [y_1, y_2, \ldots, y_m]^T$.

Key Operations

Essential operations include:

- Matrix multiplication: $(AB)_{ij} = \sum_k a_{ik} b_{kj}$;

- Vector dot product: $a \cdot b = \sum_i a_i b_i$;

- Matrix transpose: $(A^T)_{ij} = A_{ji}$.

16.2.3   Calculus for Optimization

Calculus concepts are crucial for training supervised learning models through optimization.

Derivatives and Gradients

The gradient indicates the direction of steepest increase:

$$\nabla f = \left[\frac{\partial f}{\partial x_1}, \frac{\partial f}{\partial x_2}, \ldots, \frac{\partial f}{\partial x_n} \right]^T.$$

Gradient Descent

The fundamental optimization algorithm updates parameters iteratively:

$$\theta_{t+1} = \theta_t - \alpha \nabla L(\theta_t),$$

where:

- θ_t is the parameter vector at step t;
- α is the learning rate;
- $L(\theta)$ is the loss function.

16.2.4   Information Theory Basics

Information theory helps us measure uncertainty and information content in data.

Entropy

Entropy measures uncertainty in a probability distribution:

$$H(X) = -\sum_i p(x_i) \log p(x_i).$$

Mutual Information

Mutual information quantifies dependence between variables:

$$I(X;Y) = H(X) - H(X|Y).$$

This helps in feature selection and understanding variable relationships.

KL Divergence

Kullback-Leibler divergence measures difference between distributions:

$$D_{KL}(P \parallel Q) = \sum_i P(i) \log \frac{P(i)}{Q(i)}.$$

16.2.5   Statistical Learning Theory

Statistical learning theory provides the theoretical foundation for supervised learning.

Empirical Risk Mginimization

The empirical risk is the average loss on training data:

$$R_{emp}(f) = \frac{1}{n} \sum_{i=1}^{n} L(y_i, f(x_i)).$$

Bias-Variance Decomposition

The prediction error can be decomposed:

$$E\left[(y - \hat{f}(x))^2\right] = \mathrm{Var}(\hat{f}(x)) + \left[\mathrm{Bias}(\hat{f}(x))\right]^2 + \sigma^2.$$

This decomposition helps understand the trade-off between model complexity and generalization.

Generalization Bounds

The generalization error is bounded by the empirical error plus a complexity term:

$$R(f) \le R_{emp}(f) + \sqrt{\frac{\log(1/\delta)}{2n}},$$

where:

- $R(f)$ is the true risk;
- n is the sample size;
- δ is the confidence parameter.

These mathematical foundations provide the tools needed to understand and implement supervised learning algorithms. While the mathematics can be complex, the underlying principles guide us in developing effective learning systems. The next sections will build upon these foundations to explore specific algorithms and their implementations.

16.3   Classification Algorithms

Classification algorithms form the foundation of supervised learning applications [1, 2]. This section explores key classification algorithms and their practical implementation.

16.3.1   Logistic Regression

Logistic regression models the probability of class membership using the logistic function [1, 2].

Theory

The logistic function maps any real value to the range [0,1]:

$$P(Y = 1|X) = \frac{1}{1 + e^{-\left(\beta_0 + \beta^T X\right)}}.$$

The decision boundary is linear in the feature space. The model parameters β are learned by maximizing the likelihood function:

$$L(\beta) = \prod_{i=1}^{n} P(Y = y_i|;x_i|;\beta).$$

Implementation

```python
from sklearn.linear_model import LogisticRegression
# Create and train the model
model = LogisticRegression()
model.fit(X_train, y_train)
# Make predictions
y_pred = model.predict(X_test)
# Get probability estimates
prob_estimates = model.predict_proba(X_test)
```

16.3.2   Support Vector Machines (SVM)

SVMs find the optimal hyperplane that maximizes the margin between classes.

Theory

The SVM optimization problem is:

$$\min_{\mathbf{w},b} \frac{1}{2}|\mathbf{w}|^2,$$

$$\text{subject to}: y_i\left(\mathbf{w}^T\mathbf{x}_i + b\right) \geq 1 \text{ for all } i.$$

For non-linear classification, the kernel trick transforms the feature space:

$$K\left(\mathbf{x},\mathbf{x}'\right) = \phi\left(\mathbf{x}\right)^T \phi\left(\mathbf{x}'\right).$$

Implementation

```python
from sklearn.svm import SVC
# Create SVM classifier with RBF kernel
svm = SVC(kernel='rbf', C=1.0)
svm.fit(X_train, y_train)
```

```
# Make predictions
predictions = svm.predict(X_test)
```

16.3.3   Decision Trees

Decision trees partition the feature space through recursive binary splitting.

Theory

At each node, the split criterion maximizes information gain:

$$\mathrm{IG}\left(D_p, f\right) = H\left(D_p\right) - \sum_{j=1}^{m} \frac{\left|D_j\right|}{\left|D_p\right|} H\left(D_j\right),$$

where:

- $H(D)$ is the entropy of dataset D;
- D_p is the parent node dataset;
- D_j are the child node datasets.

Implementation

```
from sklearn.tree import DecisionTreeClassifier
# Create and train decision tree
tree = DecisionTreeClassifier(max_depth=5)
tree.fit(X_train, y_train)
# Visualize the tree
from sklearn.tree import plot_tree
plot_tree(tree, feature_names=feature_names)
```

16.3.4   Random Forests

Random forests combine multiple decision trees through ensemble learning.

Theory

Each tree is trained on a bootstrap sample of the data. The final prediction is made by majority voting:

$$\hat{y} = \mathrm{mode}\left(f_1\left(x\right), f_2\left(x\right), \dots f_m\left(x\right)\right),$$

where $f_i(x)$ is the prediction of the i-th tree.

Random feature selection at each split improves diversity:

- For each split, consider only a random subset of features;
- Typically $\sqrt{p}$ features where p is the total number of features.

Implementation

```
from sklearn.ensemble import RandomForestClassifier
# Create and train random forest
rf = RandomForestClassifier(n_estimators=100)
rf.fit(X_train, y_train)
# Feature importance
importances = rf.feature_importances_
```

16.3.5 Practical Considerations

Model Selection

Choose the appropriate algorithm based on:

- Dataset size and dimensionality;
- Feature characteristics;
- Required interpretability;
- Computational constraints.

Parameter Tuning

Key parameters for each algorithm:

- Logistic Regression: regularization strength;
- SVM: kernel type, C parameter;
- Decision Trees: maximum depth, minimum samples per leaf;
- Random Forests: number of trees, maximum features.

Example: Complete Classification Pipeline

```
from sklearn.pipeline import Pipeline
from sklearn.preprocessing import StandardScaler
```

```
# Create classification pipeline
pipeline = Pipeline([
    ('scaler', StandardScaler()),
    ('classifier', RandomForestClassifier())
])
# Train pipeline
pipeline.fit(X_train, y_train)
# Evaluate
from sklearn.metrics import classification_report
print(classification_report(y_test, pipeline.predict(X_test)))
```

Classification algorithms offer different approaches to supervised learning problems. Each algorithm has unique strengths and trade-offs. Understanding these differences helps in selecting the most appropriate method for specific applications. The next section will explore regression algorithms for continuous output prediction.

16.4   Regression Algorithms

Regression algorithms predict continuous numerical values [1, 2]. This section explores fundamental regression techniques and their implementation in Python.

16.4.1   Linear Regression

Linear regression models the relationship between features and target variables using a linear equation.

Theory

The linear regression model is expressed as:

$$y = \beta_0 + \beta^T X + \epsilon,$$

where:

- β_0 is the intercept;
- β is the coefficient vector;
- X is the feature vector;
- ϵ is the error term.

The parameters are estimated by minimizing the sum of squared errors:

$$\min_{\beta_0,\beta} \sum_{i=1}^{n} \left(y_i - \beta_0 - \beta^T x_i \right)^2 .$$

Implementation

```python
from sklearn.linear_model import LinearRegression
# Create and train model
model = LinearRegression()
model.fit(X_train, y_train)
# Make predictions
y_pred = model.predict(X_test)
# Access coefficients
print("Intercept:", model.intercept_)
print("Coefficients:", model.coef_)
```

16.4.2   Polynomial Regression

Polynomial regression extends linear regression to model nonlinear relationships.

Theory

The polynomial regression model of degree d is:

$$y = \beta_0 + \beta_1 x + \beta_2 x^2 + \ldots + \beta_d x^d + \epsilon.$$

This can be viewed as linear regression with transformed features.

Implementation

```python
from sklearn.preprocessing import PolynomialFeatures
from sklearn.pipeline import make_pipeline
# Create polynomial regression model
degree = 2
polyreg = make_pipeline(
    PolynomialFeatures(degree),
    LinearRegression()
)
# Train model
polyreg.fit(X_train, y_train)
```

16.4.3   Regularized Regression

Regularization prevents overfitting by adding penalty terms to the objective function.

Ridge Regression (L2 Regularization)

Ridge regression adds a squared magnitude constraint:

$$\min_{\beta_0,\beta} \sum_{i=1}^{n} \left(y_i - \beta_0 - \beta^T x_i \right)^2 + \lambda \left| \beta \right|_2^2,$$

where λ controls regularization strength.

```
from sklearn.linear_model import Ridge
ridge = Ridge(alpha=1.0)  # alpha is λ
ridge.fit(X_train, y_train)
```

Lasso Regression (L1 Regularization)

Lasso uses absolute value regularization:

$$\min_{\beta_0,\beta} \sum_{i=1}^{n} \left(y_i - \beta_0 - \beta^T x_i \right)^2 + \lambda \left| \beta \right|_1.$$

This can lead to sparse solutions (feature selection).

```
from sklearn.linear_model import Lasso
lasso = Lasso(alpha=1.0)
lasso.fit(X_train, y_train)
```

16.4.4   Practical Applications

House Price Prediction

```
# Complete regression pipeline
from sklearn.pipeline import Pipeline
from sklearn.preprocessing import StandardScaler
pipeline = Pipeline([
    ('scaler', StandardScaler()),
    ('poly', PolynomialFeatures(degree=2)),
    ('regression', Ridge(alpha=0.1))
```

```
])
# Train and evaluate
pipeline.fit(X_train, y_train)
mse = mean_squared_error(y_test, pipeline.predict(X_test))
```

Time Series Forecasting

```
# Time series regression example
def create_time_features(df):
    df['hour'] = df.index.hour
    df['day'] = df.index.day
    df['month'] = df.index.month
    return df
# Create and train model
time_model = Ridge(alpha=0.1)
time_model.fit(X_time_features, y_values)
```

Demand Prediction

```
# Multiple feature regression
demand_model = Pipeline([
    ('scaler', StandardScaler()),
    ('regression', LinearRegression())
])
# Add cross-validation
from sklearn.model_selection import cross_val_score
scores = cross_val_score(demand_model, X, y, cv=5)
```

16.4.5   Implementation Best Practices

Feature Preprocessing

```
# Standard preprocessing pipeline
preprocess = Pipeline([
    ('imputer', SimpleImputer(strategy='median')),
    ('scaler', StandardScaler()),
    ('poly', PolynomialFeatures(degree=2))
])
```

Model Evaluation

```python
from sklearn.metrics import r2_score, mean_squared_error
import numpy as np
# Calculate metrics
r2 = r2_score(y_test, y_pred)
rmse = np.sqrt(mean_squared_error(y_test, y_pred))
```

Hyperparameter Tuning

```python
from sklearn.model_selection import GridSearchCV
# Parameter grid
param_grid = {
    'alpha': [0.001, 0.01, 0.1, 1.0, 10.0],
    'fit_intercept': [True, False]
}
# Grid search
grid_search = GridSearchCV(Ridge(), param_grid, cv=5)
grid_search.fit(X_train, y_train)
```

Regression algorithms provide powerful tools for predicting continuous values. The choice between different techniques depends on the specific problem characteristics:

- Linear regression for simple linear relationships;
- Polynomial regression for nonlinear patterns;
- Ridge and Lasso for handling overfitting and feature selection.

Understanding these algorithms and their implementations enables effective solution development for real-world regression problems.

16.5   Model Evaluation and Validation

Model evaluation and validation are crucial steps in supervised learning [1, 2]. They help us understand model performance and ensure generalization to new data.

16.5.1   Performance Metrics for Classification

Classification models require specific metrics to assess their performance accurately.

Accuracy

The simplest metric is accuracy, representing the proportion of correct predictions:

$$\text{Accuracy} = \frac{\text{Number of correct predictions}}{\text{Total number of predictions}}.$$

```
from sklearn.metrics import accuracy_score
accuracy = accuracy_score(y_true, y_pred)
```

Precision and Recall

These metrics provide deeper insight into model performance:

$$\text{Precision} = \frac{\text{True Positives}}{\text{True Positives} + \text{False Positives}},$$

$$\text{Recall} = \frac{\text{True Positives}}{\text{True Positives} + \text{False Negatives}}.$$

```
from sklearn.metrics import precision_score, recall_score
precision = precision_score(y_true, y_pred)
recall = recall_score(y_true, y_pred)
```

F1 Score

The F1 score combines precision and recall:

$$F1 = 2 \times \frac{\text{Precision} \times \text{Recall}}{\text{Precision} + \text{Recall}}.$$

```
from sklearn.metrics import f1_score
f1 = f1_score(y_true, y_pred)
```

ROC and AUC

The Receiver Operating Characteristic (ROC) curve plots true positive rate against false positive rate:

```
from sklearn.metrics import roc_curve, auc
fpr, tpr, _ = roc_curve(y_true, y_pred_proba)
roc_auc = auc(fpr, tpr)
```

16.5.2   Performance Metrics for Regression

Regression models require different metrics focused on prediction error.

Mean Squared Error (MSE)

MSE measures the average squared difference between predictions and true values:

$$\text{MSE} = \frac{1}{n}\sum_{i=1}^{n}\left(y_i - \hat{y}_i\right)^2 .$$

```
from sklearn.metrics import mean_squared_error
mse = mean_squared_error(y_true, y_pred)
```

Root Mean Squared Error (RMSE)

RMSE provides error in the same units as the target variable:

$$\text{RMSE} = \sqrt{\frac{1}{n}\sum_{i=1}^{n}\left(y_i - \hat{y}_i\right)^2} .$$

```
rmse = np.sqrt(mean_squared_error(y_true, y_pred))
```

R-Squared Score

R-squared indicates the proportion of variance explained by the model:

$$R^2 = 1 - \frac{\sum_{i=1}^{n}\left(y_i - \hat{y}i\right)^2}{\sum i = 1^n \left(y_i - \bar{y}\right)^2} .$$

```
from sklearn.metrics import r2_score
r2 = r2_score(y_true, y_pred)
```

16.5.3   Cross-validation Techniques

Cross-validation helps assess model performance on unseen data.

K-Fold Cross-validation

The dataset is split into k folds:

```python
from sklearn.model_selection import KFold
kf = KFold(n_splits=5, shuffle=True)
for train_index, val_index in kf.split(X):
    X_train, X_val = X[train_index], X[val_index]
    y_train, y_val = y[train_index], y[val_index]
```

Stratified K-Fold

Maintains class distribution in each fold:

```python
from sklearn.model_selection import StratifiedKFold
skf = StratifiedKFold(n_splits=5, shuffle=True)
```

Leave-One-Out Cross-validation

Uses each point as validation once:

```python
from sklearn.model_selection import LeaveOneOut
loo = LeaveOneOut()
```

16.5.4   Training/Validation/Test Splits

Proper data splitting is essential for reliable model evaluation.

Basic Split

The standard approach divides data into three sets:

```python
from sklearn.model_selection import train_test_split
# First split: training and temporary testing
X_train, X_temp, y_train, y_temp = train_test_split(
    X, y, test_size=0.3, random_state=42
```

```
)
# Second split: validation and testing
X_val, X_test, y_val, y_test = train_test_split(
    X_temp, y_temp, test_size=0.5, random_state=42
)
```

Stratified Split

Maintains class distribution:

```
X_train, X_test, y_train, y_test = train_test_split(
    X, y, test_size=0.2, stratify=y, random_state=42
)
```

16.5.5   Model Selection Strategies

Effective model selection combines different evaluation techniques.

Grid Search

Systematically works through multiple parameter combinations:

```
from sklearn.model_selection import GridSearchCV
param_grid = {
    'max_depth': [3, 5, 7],
    'min_samples_split': [2, 5, 10]
}
grid_search = GridSearchCV(
    estimator=model,
    param_grid=param_grid,
    cv=5,
    scoring='accuracy'
)
```

Random Search

Samples parameter combinations randomly:

```
from sklearn.model_selection import RandomizedSearchCV
random_search = RandomizedSearchCV(
    estimator=model,
```

```
    param_distributions=param_grid,
    n_iter=100,
    cv=5
)
```

Pipeline Validation

Combines preprocessing and model selection:

```python
from sklearn.pipeline import Pipeline
pipeline = Pipeline([
    ('scaler', StandardScaler()),
    ('classifier', RandomForestClassifier())
])
# Validate entire pipeline
cv_scores = cross_val_score(pipeline, X, y, cv=5)
```

Model evaluation and validation are critical for developing reliable supervised learning solutions. The choice of metrics and validation techniques should align with the problem requirements and data characteristics. Proper evaluation ensures that models will perform well on new, unseen data in real-world applications.

16.6   Practice and Exercises

This section provides hands-on exercises to reinforce the concepts of super-vised learning through practical implementation. Each exercise builds upon previous knowledge while introducing new challenges.

All exercises are available in an interactive Python notebook that can be accessed using the QR code below or at: https://colab.research.google.com/drive/1zI8RIZCO 7f46wtaulWpBu3TDwGjTBuvO

The exercises focus on real-world scenarios and common implementation tasks you will encounter in machine learning projects. Solutions to the exercises are available to instructors through the publisher's resources.

16.6.1   Practice and Programming Exercises

Exercise 1: Basic Binary Classification

Learning Objectives:
- Implement a basic supervised learning pipeline;
- Understand data splitting and model evaluation;
- Visualize classification results;
- Interpret model performance metrics;

  Tasks:

1. Generate a synthetic classification dataset with two features;
2. Split data into training and testing sets;
3. Train a logistic regression classifier;
4. Evaluate model performance using accuracy and confusion matrix;
5. Create visualizations of classification results;
6. Analyze and interpret the results;

Expected Output:
- Classification accuracy above 0.80;
- Clear visualization of decision boundaries;
- Properly formatted confusion matrix;
- Written analysis of model performance.

Exercise 2: Feature Engineering and Preprocessing

Learning Objectives:
- Design robust preprocessing pipelines;
- Handle multiple feature types;
- Implement feature scaling;
- Understand feature importance analysis;

  Tasks:

1. Create a dataset with mixed feature types (numerical and categorical);
2. Implement a preprocessing pipeline using ColumnTransformer;
3. Apply appropriate scaling and encoding techniques;
4. Train a model using the preprocessed features;
5. Analyze feature importance;

6. Compare model performance with and without preprocessing;

Expected Output:
- Functioning preprocessing pipeline;
- Feature importance visualization;
- Performance comparison metrics;
- Documentation of preprocessing steps.

Exercise 3: Data Preprocessing and Missing Values

Learning Objectives:
- Handle missing data effectively;
- Implement data transformation techniques;
- Visualize preprocessing effects;
- Create robust data pipelines;

Tasks:

1. Create a dataset with missing values and categorical features;
2. Implement multiple imputation strategies;
3. Create visualizations of data before and after preprocessing;
4. Compare the effectiveness of different preprocessing approaches;
5. Document the impact of preprocessing choices;

Expected Output:
- Preprocessed dataset visualization;
- Comparison of imputation strategies;
- Analysis of preprocessing impact;
- Implementation documentation.

Exercise 4: Model Evaluation and Metrics

Learning Objectives:
- Implement comprehensive model evaluation;
- Understand different performance metrics;
- Create performance visualizations;
- Compare classification and regression metrics;

Tasks:

1. Train both classification and regression models;
2. Implement multiple evaluation metrics;
3. Create visualization of model performance;
4. Compare different evaluation approaches;
5. Analyze trade-offs between metrics;

Expected Output:
- Complete performance analysis;
- Metric comparison visualizations;
- Written evaluation report;
- Implementation of multiple metrics.

Exercise 5: Regularization and Model Complexity

Learning Objectives:
- Understand overfitting and underfitting;
- Implement regularization techniques;
- Evaluate model complexity effects;
- Compare model variants;

Tasks:

1. Create models with varying complexity levels;
2. Implement different regularization approaches;
3. Compare model performance;
4. Visualize the impact of regularization;
5. Document complexity trade-offs;

Expected Output:
- Performance comparison of model variants;
- Visualization of regularization effects;
- Analysis of complexity impact;
- Implementation documentation.

16.6.2   Self-Assessment Questions

Before attempting these questions, review the chapter material thoroughly. Take time to consider each question carefully. These questions will help evaluate your understanding of supervised learning concepts and implementations.

Multiple Choice Questions

1. Which statement about supervised learning is correct?

    (a) It requires no labeled training data
    (b) The model learns patterns from labeled examples
    (c) It works only with continuous variables
    (d) Labels are generated automatically

2. What is the primary difference between classification and regression?

   (a) Classification uses neural networks, regression uses linear models
   (b) Classification predicts discrete classes, regression predicts continuous values
   (c) Classification is supervised, regression is unsupervised
   (d) Classification works with small datasets, regression with large ones

3. When implementing logistic regression, which preprocessing step is most important?

   (a) One-hot encoding all features
   (b) Feature scaling
   (c) Removing all outliers
   (d) Converting all features to integers

4. What does high bias in a model indicate?

   (a) The model is too complex
   (b) The model underfits the data
   (c) The model has too many features
   (d) The model is overfitting

5. Which metric is most appropriate for an imbalanced classification problem?

   (a) Accuracy
   (b) F1-score
   (c) Mean squared error
   (d) R-squared

Critical Thinking Questions

6. Your model achieves 99% accuracy on the training set but 85% on the test set. What might be happening and how would you address it?
7. When would you choose a decision tree over logistic regression? Consider interpretability, data characteristics, and model assumptions.
8. How does feature scaling affect different types of models? Compare its impact on logistic regression versus decision trees.
9. What are the trade-offs between model complexity and generalization? Use examples from the chapter.
10. Explain how cross-validation helps in assessing model performance. Why is it better than a single train-test split?

Practical Application Questions

11. You are working with a dataset that has missing values and categorical features. Design a preprocessing pipeline to handle these challenges.

12. Compare two different ways to handle categorical variables in a supervised learning model. When would you use each approach?
13. You need to predict house prices. What type of supervised learning would you use and why? What evaluation metrics would be appropriate?
14. How would you determine if your features are contributing meaningfully to the model's predictions?
15. Design a strategy to prevent overfitting in a supervised learning model while maintaining good predictive performance.

Answer Key

1. (b) Models in supervised learning learn patterns from labeled examples to make predictions.
2. (b) Classification predicts discrete categories while regression predicts continuous numerical values.
3. (b) Feature scaling is crucial for logistic regression as it uses gradient-based optimization.
4. (b) High bias indicates underfitting, where the model is too simple to capture the data patterns.
5. (b) F1-score balances precision and recall, making it suitable for imbalanced datasets.
6. This indicates overfitting. Solutions include:

   - Adding regularization
   - Reducing model complexity
   - Collecting more training data
   - Using cross-validation

7. Decision trees are preferable when:

   - Interpretability is crucial
   - Data has non-linear relationships
   - Features are on different scales
   - Missing values are present

8. Feature scaling:

   - Essential for logistic regression and neural networks
   - Not required for decision trees
   - Affects gradient-based optimization
   - Improves convergence speed

9. Trade-offs include:

   - Simple models may underfit
   - Complex models may overfit
   - Need to balance based on data size
   - Consider validation performance

10. Cross-validation:

   - Provides more reliable performance estimates
   - Uses data more efficiently
   - Reduces dependence on single splits
   - Helps detect overfitting

Questions 11–15 have multiple valid approaches. Evaluate responses based on:

- Clear problem understanding
- Logical solution approach
- Consideration of practical constraints
- Proper use of technical concepts

Focus on understanding the concepts rather than memorizing specific answers. These questions help develop practical problem-solving skills in supervised learning applications.

16.7 Conclusions and Future Directions

Supervised learning represents a powerful approach to building intelligent systems. This chapter has explored its fundamental principles, mathematical foundations, and practical implementations. The journey from basic concepts to advanced applications reveals both the power and limitations of supervised learning methods.

Our exploration covered several key aspects. First, we established the theoretical foundations including probability theory, linear algebra, and optimization principles. Second, we examined classification algorithms such as logistic regression, support vector machines, decision trees, and random forests. Third, we investigated regression techniques for predicting continuous values through linear, polynomial, and regularized models. Finally, we emphasized proper model evaluation and validation to ensure generalization.

The field of supervised learning continues to evolve rapidly. Three important future directions emerge. Automated Machine Learning advances will make supervised learning more accessible to non-experts through automated feature selection and model tuning. Interpretable Models will balance performance with transparency, addressing growing concerns about AI accountability. Edge Computing Applications will require more efficient algorithms as supervised learning moves to mobile and IoT devices.

Key takeaways include understanding the complete supervised learning pipeline, recognizing the importance of proper model evaluation, appreciating the balance between model complexity and interpretability, and developing critical thinking about algorithm selection. These fundamentals remain relevant even as specific technologies evolve.

References

1. Norvig, P.: Artificial intelligence: a modern approach. Global edition. Pearson, Boston (2021).
2. Russell, S., Norvig, P.: Artificial Intelligence: A Modern Approach. Pearson, Hoboken (2020).
3. Tyagi, K., Rane, C., Manry, M.: Chapter 1—Supervised learning. In: Pandey, R., Khatri, S.K., Singh, N. Kumar, and Verma, P. (eds.) Artificial Intelligence and Machine Learning for EDGE Computing. pp. 3–22. Academic Press (2022). https://doi.org/10.1016/B978-0-12-824054-0.00004-6.
4. Tiwari, A.: Chapter 2—Supervised learning: From theory to applications. In: Pandey, R., Khatri, S.K., Singh, N. Kumar, and Verma, P. (eds.) Artificial Intelligence and Machine Learning for EDGE Computing. pp. 23–32. Academic Press (2022). https://doi.org/10.1016/B978-0-12-824054-0.00026-5.
5. Rajoub, B.: Chapter 3—Supervised and unsupervised learning. In: Zgallai, W. (ed.) Biomedical Signal Processing and Artificial Intelligence in Healthcare. pp. 51–89. Academic Press (2020). https://doi.org/10.1016/B978-0-12-818946-7.00003-2.
6. Belyadi, H., Haghighat, A.: Chapter 5—Supervised learning. In: Belyadi, H. and Haghighat, A. (eds.) Machine Learning Guide for Oil and Gas Using Python. pp. 169–295. Gulf Professional Publishing (2021). https://doi.org/10.1016/B978-0-12-821929-4.00004-4.
7. Li, Z., Ding, Z.: Chapter 11—Discrete-time algorithms to supervised learning. In: Li, Z. and Ding, Z. (eds.) Distributed Optimization and Learning. pp. 207–229. Academic Press (2024). https://doi.org/10.1016/B978-0-44-321636-7.00020-9.
8. Morales, E.F., Escalante, H.J.: Chapter 6—A brief introduction to supervised, unsupervised, and reinforcement learning. In: Torres-García, A.A., Reyes-García, C.A., Villaseñor-Pineda, L., and Mendoza-Montoya, O. (eds.) Biosignal Processing and Classification Using Computational Learning and Intelligence. pp. 111–129. Academic Press (2022). https://doi.org/10.1016/B978-0-12-820125-1.00017-8.

GPSR Compliance
The European Union's (EU) General Product Safety Regulation (GPSR) is a set
of rules that requires consumer products to be safe and our obligations to
ensure this.

If you have any concerns about our products, you can contact us on

ProductSafety@springernature.com

In case Publisher is established outside the EU, the EU authorized
representative is:

Springer Nature Customer Service Center GmbH
Europaplatz 3
69115 Heidelberg, Germany